Colloidal Ceramic Processing of Nano-, Micro-, and Macro-Particulate Systems

T0329145

Other books published by The American Ceramic Society

Introduction to Drying of Ceramics
Denis A. Brosnan and Gilbert C. Robinson
©2003, ISBN 1-57498-046-7

Introduction to Ceramic Engineering Design
Edited by David E. Clark, Diane C. Folz, and Thomas D. McGee
©2002, ISBN 1-57498-131-5

Ceramic Nanomaterials and Nanotechnology II (Ceramic Transactions Volumer 148)
Mark R. De Guire, Michael Z. Hu, Yury Gogotsi, and Song Wei Lu, Editors
©2004, ISBN 1-57498-203-6

Ceramic Nanomaterials and Nanotechnology (Ceramic Transactions Volume 137)
Edited by Michael Z. Hu and Mark R. De Guire
©2002, ISBN 1-57498-152-8

Progress in Nanotechnology
©2002, ISBN 1-57498-168-4

Innovative Processing and Synthesis of Ceramics, Glasses, and Composites VII (Ceramic Transactions Volume 154)
Edited by Bansal and J.P. Singh and Narottam P.
©2004, ISBN 1-57498-208-7

Innovative Processing and Synthesis of Ceramics, Glasses, and Composites VI (Ceramic Transactions Volume 135)
Edited by Narottam P. Bansal and J.P. Singh
©2002, ISBN 1-57498-150-1

The Magic of Ceramics
David W. Richerson
©2000, ISBN 1-57498-050-5

For information on ordering titles published by The American Ceramic Society, or to request a publications catalog, please contact our Customer Service Department at:

Customer Service Department
735 Ceramic Place
Westerville, OH 43081, USA
614-794-5890 (phone)
614-794-5892 (fax)
info@ceramics.org

Visit our on-line book catalog at www.ceramics.org.

Volume 152

Colloidal Ceramic Processing of Nano-, Micro-, and Macro- Particulate Systems

Proceedings of the Colloidal Ceramic Processing: Nano-, Micro-, and Macro-Particulate Systems held at the 105th Annual Meeting of The American Ceramic Society, April 27–30, 2003 in Nashville, Tennessee

Edited by

Wei-Heng Shih
Drexel University

Yoshihiro Hirata
Kagoshima University

William Carty
Alfred University

Published by
The American Ceramic Society
735 Ceramic Place
Westerville, Ohio 43081
www.ceramics.org

Proceedings of the Colloidal Ceramic Processing: Nano-, Micro-, and Macro-Particulate Systems held at the 105th Annual Meeting of The American Ceramic Society, April 27–30, 2003 in Nashville, Tennessee

COVER PHOTO: A single silicon nitride spray-dried particle sits on top of a commercial tip for atomic force microscopy evaluation is courtesy of H. Kamiya, S. Matsui and T. Kakui and appears as part of figure 2 in their paper "Analysis of Action Mechanism of Anionic Polymer Dispersant with Different Molecular Structure in Dense Silicon Nitride Suspension by using Colloidal Probe AFM" which begins on page 83.

For information on ordering titles published by The American Ceramic Society, or to request a publications catalog, please call 614-794-5890.

4 3 2 1–07 06 05 04

ISSN 1042-1122
ISBN 1-57498-211-7

Contents

Preface. vii

Dispersibility of Nanometer-Sized Ceria Particles I
Y. Hirata, H. Takahashi, H. Shimazu, and S. Sameshima

Glass-Ceramic Thin Films by Sol-Gel Process for
Electronic Application II
K. Saegusa

Preparation of Highly Dispersed Ultra-Fine Barium Titanate
Powder by Using Acrylic Oligomer with High Density of
Hydrophilic Group 27
Y. Yonemochi, Y. Iida, K. Ogino, H. Kamiya, K. Gomi, and K. Tanaka

The Thermal Stability and Structural Properties Evolution of
Cured and Non-Cured ZrO_2 and ZrO_2-SiO_2 Powders 37
Q. Zhao, Wan Y. Shih, and W.Y.-H. Shih

Microporous Silica Modified with Alumina as
CO_2/N_2 Separators 47
T. Patil, Q. Zhao, R. Mutharasan, W. Shih, and W.-H. Shih

Study of Mechanism of Pyrochlore-Free PMN-PT Powder
using a Coating Method. 55
H. Gu, W. Shih, and W.-H. Shih

Micromechanical Testing of Two-Dimensional Aggregated
Colloids. 65
S. Promkotra and K.T. Miller

Rheology of Ceramic Slurries for Electro-Deposition in
Rapid Prototyping Applications 75
N. Manjooran, S. Lee, G. Pyrgiotakis, and W. Sigmund

Analysis of Action Mechanism of Anionic Polymer Dispersant
with Different Molecular Structure in Dense Ceramic
Suspension by using Colloidal Probe AFM 83
H. Kamiya, S. Matsui, and T. Kakui

Aqueous Processing of WC-Co Powders: Suspension
Preparation and Granule Properties 93
K.M. Andersson and L. Bergstrom

Colloidal Processing and Liquid Phase Sintering of
SiC-Al$_2$O$_3$-Y^{3+} Ions System. 109
N. Hidaka and Y. Hirata

Colloidal Processing of SiC with 700 MPa of
Flexural Strength 119
S. Tabata and Y. Hirata

Adsorption of Poly(acrylic acid) on Commercial Ball Clay. . . . 129
U. Kim, B. Schulz, and W. Carty

Analysis of Thickness Control Variables in Tape Casting
Part II: The Effect of Blade Gap 139
M. Gifford, E. Twiname, and R. Mistler

Index 147

Preface

This volume of Ceramic Transactions consists of papers presented at the International Symposium on Colloidal Ceramic Processing which was held during the 105th Annual Meeting of The American Ceramic Society in Nashville, Tennessee, April 27-30, 2003. The papers summarize the recent developments in colloidal ceramic processing and cover a broad range of subjects including nano-, micro-, and macro-particulate systems. The editors would like to thank the ACerS Basic Science Division for their financial support for this symposium.

Wei-Heng Shih

Yoshihiro Hirata

William Carty

DISPERSIBILITY OF NANOMETER-SIZED CERIA PARTICLES

Yoshihiro Hirata, Hideyuki Takahashi, Hisanori Shimazu
and Soichiro Sameshima
Department of Advanced Nanostructured Materials Science and Technology,
Graduate School of Science and Engineering, Kagoshima University
1-21-40 Korimoto, Kagoshima 890-0065, Japan

ABSTRACT

Nanometer-sized ceria particles (average diameter of 15 nm, calculated from specific surface area) were dispersed at 2 vol% solid in aqueous solutions with glutamic acid and lysine of 0.05-5.0 mass% against ceria at pH 2.0-10.0. The ceria particles without amino acid were charged in the range from 12 to −34 mV at pH 2-10 and no effect of charge on the dispersibility of the powder was seen. A small amount of the amino acid was adsorbed on the positively charged ceria particles at pH 2.0 and increased the zeta potential. Addition of glutamic acid and lysine enhanced the dispersibility of the nanometer-sized ceria particles by electrosteric stabilization.

INTRODUCTION

Colloidal processing comprises of the dispersion of fine ceramic particles in a liquid medium and subsequent consolidation of the colloidal particles [1-4]. This assists in producing a uniform microstructure and high packing density for the powder compact. The dispersibility of the colloidal suspension affects both the sintering behavior and resultant properties of the densified ceramics. Dispersants such as polyacrylic acid [5, 6], ammonium salt of polyacrylic acid [7], polymethacrylic acid [8], ammonium polymethacrylate [9,10] and phosphate ester [11] are added to prepare a stable suspension with high content of ceramic powder and suitable viscosity. When the size of colloidal particles becomes smaller, controlling the properties of the colloidal suspension becomes difficult. This is mainly due to the increased interparticle interaction.i.e., the high viscosity at a low solid content, the low dispersibility of colloidal particles due to increased attractive force and the large amount of dispersant required to cover the high specific surface area of the colloidal particles [11-16]. A method to solve these problems is by finding a low molecular weight dispersant with good dispersibility. The following criteria were looked into a possible dispersant for the nanometer-sized particles : (1) the size ratio of the established dispersant to the submicrometer-sized particles in a well-dispersed suspension should be kept in a nanometer-sized particle suspension (steric stabilization effect) [7, 8, 17] and (2) the low molecular weight dispersant should be highly charged compared to the nanometer-sized particles to give a strong electrostatic stabilization effect [18, 19]. Thus, amino acid with the two types of ionizable groups (carboxyl and amino groups) can be a good choice as the dispersant. These ionized groups provide the adsorbtion onto particle surfaces and the electrostatic repulsion, depending on the pH of the suspension. Here, a nanometer-sized ceria powder was dispersed in an aqueous solution with glutamic acid and lysine to investigate their interactions.

EXPERIMENTAL PROCEDURE
Ceria powder

Table I. Heating temperature and specific surface area of ceria powder.

Drying or calcination temperature (°C)	Specific surface area (m²·g⁻¹)	Diameter of equivalent spherical particle (nm)
200	111.1	10.4
500	76.6	15.0
600	38.0	30.2
700	0.62	1871

A fine ceria powder was collected by drying a commercial ceria suspension (Nyacol Nano Technologies, Inc., MA, USA) using rotary evaporator. Figure 1 (a) shows the infrared spectrum of the dried ceria powder. HNO_3 and CH_3COOH in the starting suspension remained on the surface of dried ceria powder. Thermo Gravimetric Differential Thermal Analysis (TG-DTA) of the dried ceria powder showed 9.8 % of weight loss upon heating to 1000 °C. Figure 1 (b) shows the infrared spectrum of ceria powder calcined at 500 °C for 1 h. The remaining HNO_3 and CH_3COOH were removed by calcination. Table I indicates the influence of heating temperature on the specific surface

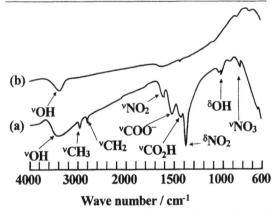

Fig. 1 Infrared spectra of (a) ceria powder and (b) ceria powder calcined at 500°C.

area. The specific surface area decreased drastically for the ceria calcined at 700 °C. The diameter (D) of equivalent spherical ceria particle was determined from the specific surface area (S) and powder density (ρ) using the relationship of D=6/ρS. For these experiments, 15 nm ceria powder calcined at 500 °C was used.

Dispersant

In Table II the chemical properties of glutamic acid and lysine used are shown [20]. Glutamic acid and lysine have the molecular length of 1.1– 1.2 nm and provide a size ratio of 0.07– 0.08 to the diameter of ceria particle. The isoelectric point for equal charge of NH_3^+ and COO^- of glutamic acid and lysine was pH 3.22 and 9.47, respectively. The dissociation of glutamic acid (GH_2) is seen in Eqs. (1) (2) and (3). Ka represents the equilibrium contant.

Table II. Chemical properties of glutamic acid and lysine

Property	Glutamic acid	Lysine
Chemical formula	$HOOC(CH_2)_2CHCOOH$ NH_2	$H_2N(CH_2)_4CHCOOH$ NH_2
Purity (mass%)	99	99
Molecular weight	147.13	146.19
Molecular length (nm)	1.1	1.2
Isoelectric point	3.22	9.47
Solubility	water soluble	water soluble
pKa1	2.19	2.20
pKa2	4.25	8.90
pKa3	9.67	10.28

$$GH_3^+ \; \underset{\longleftarrow}{\overset{Ka_1}{\longrightarrow}} \; GH_2 + H^+ \qquad (1)$$

$$GH_2 \; \underset{\longleftarrow}{\overset{Ka_2}{\longrightarrow}} \; GH^- + H^+ \qquad (2)$$

$$GH^- \; \underset{\longleftarrow}{\overset{Ka_3}{\longrightarrow}} \; G^{2-} + H^+ \qquad (3)$$

From Eqs. (1)-(3), the dissociated fraction (α) of GH_3^+ to the total concentration of neutral and ionized glutamic acid is related to Ka and $[H^+]$ by Eq. (4).

$$\alpha_{GH_3^+} = \frac{[GH_3^+]}{[GH_3^+]+[GH_2]+[GH^-]+[G^{2-}]} = \frac{[H^+]^2}{[H^+]^2 + Ka_1[H^+] + Ka_1Ka_2} \qquad (4)$$

Similarly, α_{GH_2}, α_{GH^-} and $\alpha_{G^{2-}}$ are given by Eqs. (5), (6) and (7), respectively.

$$\alpha_{GH_2} = \frac{\alpha_{GH_3^+} Ka_1}{[H^+]} \tag{5}$$

$$\alpha_{GH^-} = \frac{\alpha_{GH_2^+} Ka_1 Ka_2}{[H^+]^2} \tag{6}$$

$$\alpha_{G^{2-}} = \frac{\alpha_{GH_2^+} Ka_1 Ka_2 Ka_3}{[H^+]^3} \tag{7}$$

Figure 2 shows the fractions of $\alpha_{GH_3^+}$, α_{GH_2}, α_{GH^-}, and $\alpha_{G^{2-}}$ calculated by Eqs. (4)-(7) as a function of pH. With an increase of pH, the dominant species changes as follows: $GH_3^+ \to GH_2 \to GH^- \to G^{2-}$. Similarly, the dissociation of lysine (LH) is expressed by Eqs. (8)-(10).

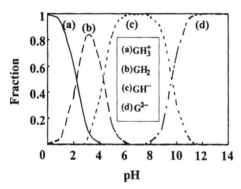

Fig. 2 Fraction of dissociated glutamic acid (GH₂) as a function of pH.

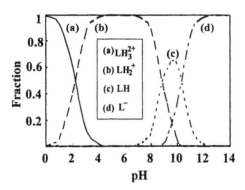

Fig. 3 Fraction of dissociated lysine (LH) as a function of pH.

$$LH_3^{2+} \underset{}{\overset{Ka_1}{\rightleftharpoons}} LH_2^+ + H^+ \tag{8}$$

$$LH_2^+ \underset{}{\overset{Ka_2}{\rightleftharpoons}} LH + H^+ \tag{9}$$

$$LH \underset{}{\overset{Ka_3}{\rightleftharpoons}} L^- + H^+ \tag{10}$$

The equilibrium constants are shown in Table II. The dissociated fractions (α) of LH_3^{2+}, LH_2^+, LH and L^- are shown in Figure 3. The dominant species changes as follow with increasing pH: $LH_3^{2+} \to LH_2^+ \to LH \to L^-$. In the pH range from 8 to 12, LH_2^+, LH and L^- coexist in a solution.

Dispersibility of ceria particles

Dispersibility of the calcined ceria particles in 2 vol% suspension with and without amino acid at pH 2-10 were evaluated by measuring the length of the phase separation into a clear solution and a concentrated suspension as a function of settling time [19]. Glutamic acid and lysine (0.05-5 mass% ceria) was added to the suspensions. The zeta potential of colloidal ceria particles with and without the amino acid was measured at pH 2-10 at a constant ionic strength of 0.01 M NH_4NO_3 (Rank Mark II, Rank Brothers Ltd., Cambrige,UK).

RESULTS AND DISCUSSION
Zeta potential of ceria powder
Figure 4 shows the zeta potential of the calcined ceria powder. The ceria particles were charged positively with 12 mV at pH 2, and charged negatively with -30 mV at pH 8. The isoelectric point was pH 3.0. Addition of 0.1 mass% glutamic acid to the ceria increased the zeta potential to 52 mV at pH 2.0.

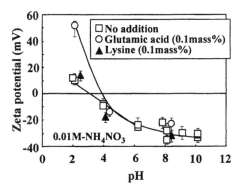

Fig. 4　Zeta potential of ceria powder in the suspensions with and without glutamic acid and lysine as a function of pH.

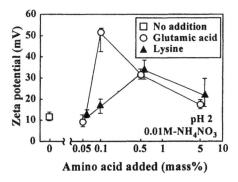

Fig. 5　Zeta potential of ceria particles in the suspensions at pH 2.0 as a function of the amount of added glutamic acid and lysine.

However, no significant influence of glutamic acid addition on the zeta potential of ceria was measured at pH 4.0 and 8.0. Figure 4 also shows the influence of lysine (0.1 mass%) on the zeta potential of ceria suspension. Figure 5 shows the relationship between the amount of glutamic acid and lysine added and the zeta potential of ceria particles at pH 2.0. The zeta potential of ceria particles showed a maximum at 0.1 mass% glutamic acid and 0.5 mass% lysine. This result indicates a strong interaction between ceria surfaces and the amino acid added.

Figure 6 shows the possible interactions between ceria surface and amino acid. In the suspension at pH 2.0, $Ce-OH_2^+$ sites are formed by the reaction, $Ce-OH + HCl \rightarrow Ce-OH_2^+ + Cl^-$. The ceria surface also reacts with OH^- ions to produce $Ce-O^-$ sites ($CeOH + OH^- \rightarrow Ce-O^- + H_2O$). The number of $Ce-OH_2^+$ sites is greater than that of $Ce-O^-$ sites at pH below 3, explaining the positively charged surface at pH 2. On the other hand, glutamic acid dissociates into 61 % GH_3^+ and 39 % GH_2 (Figure 2).

The localization of electrons in carboxyl group of GH_3^+ ion produces $O^{\delta-}$ and $C^{\delta+}$ atoms. The $O^{\delta-}$ atoms are attracted to the positively charged $Ce-OH_2^+$ site, suggesting the increased positive charge of ceria particles. The adsorption of O^- in GH_2 onto $Ce-OH_2^+$ site produces no increase of positive charge of ceria surface. However, the interaction between negative $Ce-O^-$ site and N^+ in GH_3^+ contributes to increase the surface charge to positive value because of the disappearance of negative $Ce-O^-$ sites with adsorption of GH_3^+. In the suspensions at pH 4 and 8, the number of $Ce-O^-$ sites becomes greater than that of $Ce-OH_2^+$ sites, explaining

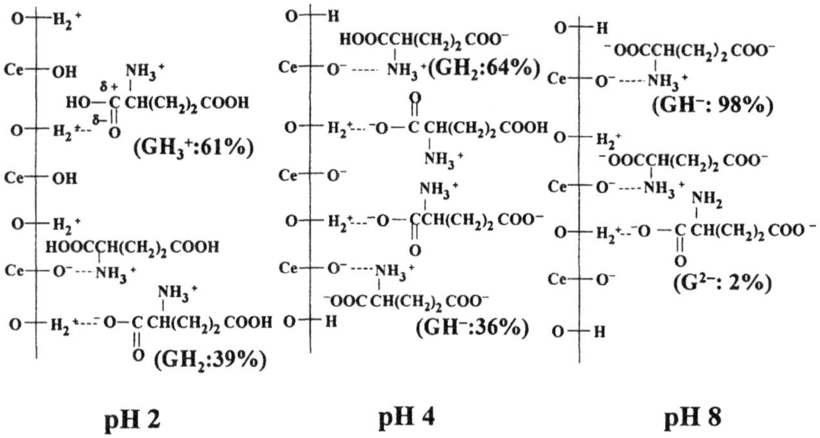

Fig. 6   Schematic illustration of possible interactions between ceria surface and glutamic acid at pH 2, 4 and 8.

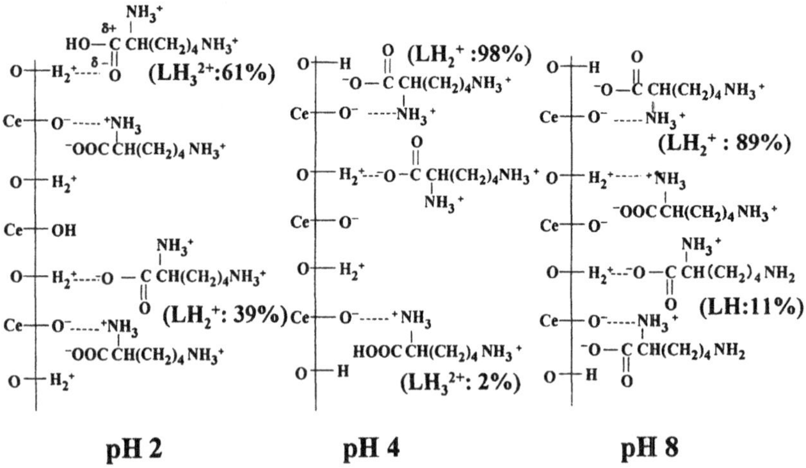

Fig. 7   Schematic illustration of possible interactions between ceria surface and lysine at pH 2, 4 and 8.

Colloidal Ceramic Processing

the negative value of zeta potential in Figure 4. The fraction of GH_2 increases to 64 % at pH 4.0. However, the adsorption of electrically neutral GH_2 onto ceria surface results in no increase of surface charge. The influence of glutamic acid addition on the zeta potential of ceria particles at pH 4.0 and 8.0 suggests that electrostatic repulsion between the negatively charged ceria surface and GH^- or G^{2-} ion suppresses the adsorption of those ions.

Figure 7 shows the possible interactions between ceria surface and lysine. In a suspension at pH 2.0, lysine dissociates into 61 % LH_3^{2+} and 39 % LH_2^+ (Figure 3). The localization of electrons in the carboxyl group of LH_3^{2+} produces $O^{\delta-}$ and $C^{\delta+}$ atoms. The interaction between $O^{\delta-}$ and $Ce-OH_2^+$, and the interaction between negative $Ce-O^-$ site and N^+ in LH_3^{2+} contribute to the increase in the surface charge (positive value). Furthermore, the adsorption of O^- in LH_2^+ onto $Ce-OH_2^+$ and the adsorption of LH_2^+ onto $Ce-O^-$ cause an increase in positive charge on the ceria surface. In a suspension at pH 4, lysine dissociates into 98 % LH_2^+. In a suspension at pH 8.0, lysine consists of 89 % LH_2^+ and 11 % LH. The interaction between $Ce-O^-$ site and N^+ in LH_3^{2+} or between $Ce-OH_2^+$ site and O^- in LH_2^+ enhances the surface charge to positive value. However, the experimental results shown in Figure 4 indicates no variation in the surface charge of ceria particles with addition of lysine at pH 4.0- 8.0.

Dispersibility of ceria particles

Figure 8 shows the length of phase separation in 2 vol% ceria suspensions without amino acid as a function of settling time. The result in Figure 8 indicates very small influence of surface charge on the dispersibility of the nanometer-sized ceria particles. The decrease of particle size is accompanied by the decrease of the repulsive energy between charged particles, resulting in the decrease of primary maximum of interaction energy (summation of repulsive energy and van der Waals energy) as a function of distance between two particles [18]. Figure 8 shows the difficulty of dispersing 15 nm ceria particles by electrostatic repulsive energy.

Figure 9 shows the effect of glutamic acid addition on the dispersibility of ceria particles at pH 2.0. Addition of glutamic acid suppressed the phase separation of ceria suspension. This result is closely related to the adsorption of glutamic acid (Figure 6) onto ceria surface and the enhancement in the zeta potential. A similar suppression effect of lysine on the phase separation was measured in the

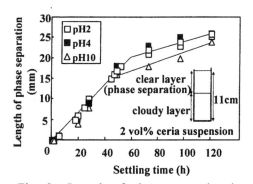

Fig. 8   Length of phase separation in the sedimentation test for ceria suspensions at pH 2, 4 and 10.

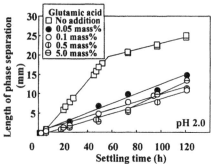

Fig. 9 Influence of glutamic acid addition on the length of phase separation in the sedimentation test for ceria suspensions at pH 2.0.

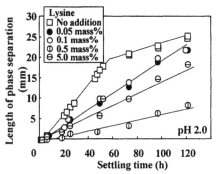

Fig. 10 Influence of lysine addition on the length of phase separation in the sedimentation test for ceria suspensions at pH 2.0.

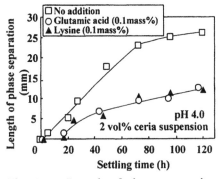

Fig. 11 Length of phase separation in ceria suspensions with and without glutamic acid at pH 4.

ceria suspension at pH 2.0 (Figure 10). Too large amount of lysine (~5 mass%) reduced both the dispersibility and zeta potential of ceria particles.

Figure 11 shows the effect of amino acid addition on the phase separation of ceria suspension at pH 4.0. Although the change in the zeta potential of ceria particles with the addition of glutamic acid and lysine was small at pH 4.0 (Figure 4), the phase separation was suppressed by the amino acid adsorbed. The result in Figure 11 suggests the steric stabilization effect of glutamic acid and lysine on the dispersion of ceria particles. Figure 12 shows the relationship between zeta potential of ceria particles and length of phase separation after 120 h of settling time at pH 2. The dependance of length of phase separation on the zeta potential was larger for lysine addition than for glutamic acid addition. This result suggests that (1) glutamic acid provides the high dispersibility due to steric stabilization effect for nanometer-sized ceria particles with low zeta potential, (2) increase of zeta potential with increasing amount of glutamic acid leads to the high dispersion of ceria particles due to electrosteric stabilization effect, and (3) the steric stabilization effect of lysine on ceria particles with low zeta potential is smaller than that of glutamic acid.

CONCLUSIONS
(1) It was difficult to control the

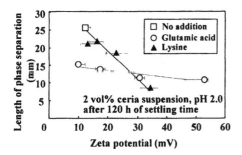

Fig. 12 Relationship between the length of phase separation after 120 h of settling time and zeta potential of ceria particles with and without glutamic acid and lysine.

dispersibility of 15 nm-size ceria particles by the surface charge in the suspensions at pH 2-10.

(2) The addition of glutamic acid and lysine of 0.05-5 mass% against ceria particles enhanced greatly the zeta potential of ceria particles to positive values at pH 2.0. The zeta potential of positively charged ceria particles showed a maximum at 0.1 and 0.5 mass% of glutamic acid and lysine, repectively.

(3) In the ceria suspensions at pH 4.0 and 8.0, very little influence of glutamic acid and lysine addition was measured on the zeta potential for the negatively charged ceria particles.

(4) Glutamic acid adsorbed provides the high dispersibility due to steric stabilization effect for nanometer-sized ceria particles with low zeta potential range. Increase of zeta potential with increasing amount of glutamic acid and lysine leads to the high dispersion of ceria particles due to electrosteric stabilization effect.

REFERENCES

[1]F. F. Lange, "Powder Processing Science and Technology for Increased Reliability," J. Am. Ceram. Soc., 72 [1], 3-15 (1989).

[2]I. A. Aksay, "Molecular and Colloidal Engineering of Ceramics," Ceram. Inter., 17, 267-274 (1991) .

[3]J. A. Lewis, "Colloidal Processing of Ceramics," J. Am. Ceram. Soc., 83 [10], 2341-2359 (2000).

[4]Y. Hirata, S. Matushita, Y. Ishihara and H. Katsuki, "Colloidal Processing and Mechanical Properties of Whisker Reinforced Mullite Matrix Composites," J. Am. Ceram. Soc., 74 [10], 2438-2442 (1991).

[5]V. A. Hackly, "Colloidal Processing of Silicon Nitride with Poly(acrylic acid): I, Adsorption and Electrostatic Interaction," J. Am. Ceram. Soc., 80 [9], 2315-2325 (1997).

[6]Y. Hirata, J. Kamikakimoto, A. Nishimoto and Y. Ishihara, "Interaction between α-Alumina Surface and Polyacrylic Acid," J. Ceram. Soc. Jpn., 100 [1], 7-12 (1992).

[7]Y. Hirata, A. Nishimoto and Y. Ishihara, "Effects of Addition of Polyacrylic Ammonium on Colloidal Processing of α-Alumina," J. Ceram. Soc. Jpn., 100 [8], 983-990 (1992).

[8]J. C. Cerasano III and I. A. Aksay, "Stability of Aqueous α-Al$_2$O$_3$ Suspensions

with Poly(methacrylic acid) Polyelectrolyte," J. Am. Ceram. Soc., 71 [4], 250-255 (1988).

[9]J-M Cho and F. Dogan, "Colloidal Processing of Lead Lanthanum Zirconate Titanate Ceramics," J. Mater. Sci., 36 [10], 2397-2403 (2001).

[10]S. Baklouti, C. Pagnoux, T. Chartier and J. F. Baumard, "Processing of Aqueous α-Al$_2$O$_3$, α-SiO$_2$, and α-SiC Suspensions with Polyelectrolytes," J. Eur. Ceram. Soc., 17 [12], 1387-1392 (1997).

[11]A. Mukherjee, B. Maiti, A. D. Sharma, R. N. Basu and H. S. Maiti, "Correlation between Slurry Rheology, Green Density and Sintered Density of Tape Cast Yttria Stabilized Zirconia," Ceram. Inter., 27 [7], 731-739(2001).

[12]H. Shan and Z. Zhang, "Slip Casting of Nanometer Sized Tetragonal Zirconia Powder," Br. Ceram. Trans., 95 [1], 35-38 (1996).

[13]A. Dietrich and A. Neubrand, "Effect of Particle Size and Molecular Weight of Polyethylenimine on Properties of Nanoparticulate Silicon Dispersions," J. Am. Ceram. Soc., 84 [4], 806-812 (2001).

[14]Y. Hirata, I. Haraguchi and Y. Ishihara, "Rheology and Consolidation of Colloidal Suspension of Ultrafine SiO$_2$-Al$_2$O$_3$ Powder," J. Ceram. Soc. Jpn., 98 [9], 951-956 (1990).

[15]Y. Hirata, I. Haraguchi and Y. Ishihara, "Particle Size Effects on Colloidal Processing of Oxide Powders," J. Mater. Res., 7 [9], 2572-2578 (1992).

[16]A. Dietrich, A. Neubrand and Y. Hirata, "Filtration Behavior of Nanoparticulate Ceria Slurries," J. Am. Ceram. Soc., 85 [11], 2719-2724 (2002).

[17]Y. Hirata, S. Tabata and J. Ideue, "Interaction in the Silicon Carbide-Polyacrylic Acid-Yttrium Ion System," J. Am. Ceram. Soc., 86 [1], 5-11 (2003).

[18]Y. Hirata, S. Nakagama and Y. Ishihara, "Calculation of Interaction Energy and Phase Diagram for Colloidal Systems," J. Ceram. Soc. Jpn., 98 [4], 316-321 (1990).

[19] I. A. Aksay and R. Kikuchi, "Structures of Colloidal Solids"; pp. 513-521 in *Science of Ceramic Chemical Processing*, Edited by L. L. Hench and D. R. Ulrich John Wiley & Sons, Inc., New York, 1986.

[20]Kagaku Binran Kisohen I (Chemical Hand Book, Basic Version I); p.432, Edited by the Chemical Society of Japan, Maruzen Co., Tokyo, 1993

GLASS-CERAMIC THIN FILMS BY SOL-GEL PROCESS FOR ELECTRONIC APPLICATION

Kunio Saegusa
Sumitomo Chemical Co. Ltd.
Tsukuba Research Laboratory
6 Kitahara
Tsukuba, 300-2436
Japan

ABSTRACT

Glass-ceramics in combination with sol-gel process is proposed as a candidate of nanomaterials with design rational for electronic and optical application. Sol-gel process, having no glass-forming step, can suppress the devitrification of glass-ceramics with high crystalline content. Various compositions of ($PbTiO_3$ (PT), or PZT)-($PbO-B_2O_3-SiO_2$) glass-ceramic thin films are prepared on Pt/Si wafer and metal (Al, Cu, Ti) foils and the basic idea to control the composition and the structure was discussed. The technique was shown useful to achieve high crystalline content thin film materials without voids or defects at a low processing temperature.

INTRODUCTION

Nanotechnology attracts considerable attention by its expected impact on our life. However, nanomaterials with well-ordered structure are not readily available and various methods are being proposed. Glass-ceramics can be the attractive nanomaterials because the structures may be controlled through nucleation and crystallization. Unfortunately, compositions of glass-ceramics have been quite limited because of devitrification during glass-formation. Several ferroelectric glass-ceramic studies in bulk materials have appeared regarding the production and properties of the materials[1-11]. The ferroelectric crystalline phase in studied materials have included PZT[1,2], $PbTiO_3$[3-8], $BaTiO_3$[9-11] and others. Typical compositions are, for example, $PZT-GeO_2$[1], $PZT-PbO-SiO_2$[2], $PbO-TiO_2-B_2O_3-BaO-SiO_2$[3,4], $PbO-TiO_2-SiO_2$[5-7], $BaO-TiO_2-Al_2O_3-SiO_2$[9,10]. For example, Haun et al.[3] prepared

0.6PZT-0.135PbO -0.27SiO$_2$ glass-ceramics from Pb$_3$O$_4$, TiO$_2$, ZrO$_2$, SiO$_2$ using a twin-roller quencher with recrystallization at 450-1200°C for 1h. The dielectric properties of the obtained glass were ε=124, tanδ=1.4%, and Pr=2.6μC/cm^2, which are significantly low value compared with pure PZT. As is evident from above, although high crystalline phase content is desirable for the higher dielectric constant it confers, this content is quite limited using the conventional glass-ceramics technique employed in previous reports. Normally, the starting oxide powders are mixed, melted and cooled down to form a glass, followed by an annealing process to allow crystallization (Fig. 1). The efficiency of TiO$_2$ and ZrO$_2$ in promoting nucleation, however, leads to devitrification when the ferroelectric phase exceeds a small fraction (Fig. 2), a fraction too small for practical applications.

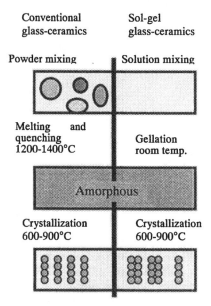

Fig.1 Schematic illustration of preparation of conventional and a new sol-gel glass-ceramics.

The purpose of high temperature melting in the conventional process is to achieve a homogeneous amorphous mixture of components. An alternative method for achieving such a molecular level mixture without resort to high temperature melting could allow glass-ceramic production while suppressing devitrification.

In this paper, an approach of utilizing a sol-gel process in combination with glass-ceramics is proposed. A sol-gel process, having no glass-forming step, can suppress the devitrification of glass-ceramics with high crystalline content. The sol-gel method is a well-known technique for preparing thin films[12-17]. These films, however, often have defects that cause short-circuits[16-17], making some applications difficult because films need to shrink in their size during sintering while rigid substrates constrain the film shrinkage. Glass-ceramic material may be the solution for this problem because it relaxes the stress during sintering and fills pores by its vitreous and pore-free nature so that they prevent thin films from cracking or forming defects.

Furthermore, the range of glass-ceramic compositions, currently limited by the difficulty of glass-formation, would be expanded. This work is the first step toward the well-ordered nanostructure such as nano-crystallites arrays in a glass matrix. This process is also applicable to thin film technology such as thin film capacitor for non-volatile memory, large-scale integration (LSI), sensors, the passive integration in PCB technology, or as a thin flexible capacitor that fits in a narrow space.[3-14)]

EXPERIMENTAL
Design Rationale for Component Levels
PT-PbO-B₂O₃ and PT-PbO-SiO₂ glass-ceramic powders and films were produced

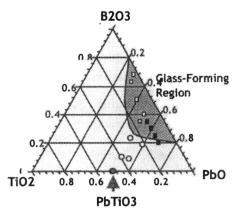

Fig.2 Composition studied in the PbO- TiO₂-B₂O₃ system: previous works ((□) glass, (■) devitrification) and this work(○).

by a sol-gel process and the crystallization behavior along with the dielectric properties was studied. PbO-TiO₂-B₂O₃ glass- ceramics is one of a few systems where the crystallization behavior dependence on its composition was reported so that we can compare the difference of the crystallization behavior between bulk materials and thin films (Fig. 2). It is also well known that PbO-B₂O₃ system forms a low melting glass in a wide composition range. Therefore, a PbO-B₂O₃ glass system was considered ideal to observe the crystallization behavior of the glass-ceramics because the liquid phase usually promotes crystallization through an enhanced mass transfer. Then a silicate system was examined because silica is expected to suppress the grain growth by its high viscosity of the glass[18,19).

The component levels within the glass-ceramics were chosen to maximize the crystal phase content. The composition of the glass-ceramics was designed to have the chemical potential of the individual components equal both in the crystal and the glass phase so that the constituents of the crystal phase do not move into

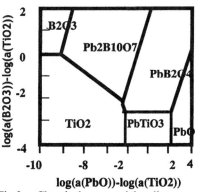

Fig.3 Chemical potential diagram of Pb-Ti-B-O system at 773K.

the glass phase. For example, the chemical potential of Pb in PT crystals should be equal to that in the glass phase at the calcination temperature when both phases are in equilibrium. Given the difficulty of accurately calculating the chemical potential of Pb in the glass phase, its potential was assumed to be equal to that in the crystal phase of the same chemical composition. The chemical potential diagram (Fig.3: $\log(a_{Pb}/a_B)$ vs $\log(a_{Pb}/a_{Ti})$ at 773 K with P_{O_2}=0.2) was drawn using a thermodynamic data base (MALT)[20).

The PbO•B$_2$O$_3$ was found to be in equilibrium with PT in case of the PbO-TiO$_2$-B$_2$O$_3$ system, and PbO·SiO$_2$ in case of the PbO-TiO$_2$-SiO$_2$ system. The composition of 2PbO•B$_2$O$_3$ was also studied due to its quasi-stable crystal phase composition on the phase diagram.

Sample Preparation

PT precursor stock solutions were prepared basically according to standard methods[12]. One mol of lead(II) acetate trihydrate was dissolved into 8 mol of 2-methoxyethanol under an N$_2$ atmosphere and then distilled at 150°C for 2 h. This distillation was repeated three times, and the lead concentration was adjusted to 1 mol/kg (Pb stock solution). Titanium tetraisopropoxide was distilled with 2-methoxyethanol to prepare the Ti stock solution (1.5 mol/kg). Acetylacetone was added to the Ti stock solution when necessary. Zirconium tetraisopropoxide for the Zr stock solution, boric acid for the B stock solution, ethyl silicate oligomer for the Si stock solution were used as starting materials. The Pb stock solution was mixed with an equimolar amount of Ti precursor stock solution, and the mixture was distilled until no distillate was observed. The concentration of this solution was adjusted to 1 mol/kg with respect to Pb (PT stock solution). The PZT stock solution was prepared in a similar manner. To achieve partial hydrolysis, H$_2$O was added with catalyst to either the PT stock solution or the PZT stock solution, which was stirred at 80°C with an [H$_2$O]/[Ti] ratio of 1.3(PT) and 4(PZT). After maturing, the solution was adjusted to 0.65M(PT) and 0.5M(PZT) with respect to Pb by reduced-pressure distillation at 80°C. The Pb stock solution, the B stock solution and the Si stock solution were mixed at 80°C for 2h to prepare the Pb-B solution (PbO•B$_2$O$_3$(PB), 2PbO•B$_2$O$_3$(2PB)), the Pb-Si solution (PbO•SiO$_2$(PS)), and the Pb-B-Si solution (3PbO•B$_2$O$_3$•SiO$_2$). These Pb-B, Pb-Si, and Pb-B-Si solutions and the PT and PZT stock solution were mixed to form the PbO-TiO$_2$-B$_2$O$_3$, PbO-TiO$_2$-SiO$_2$, PbO-TiO$_2$-ZrO$_2$-B$_2$O$_3$-SiO$_2$ solutions.

The solutions were spin-coated on a Pt(500nm)/Ti(50nm)/SiO2(400nm) /Si(100) wafer substrate and baked at 300-500°C. This procedure was repeated several times to obtain a desired thickness. Then the samples were fired at 600-900°C with a ramp rate of 600°C /min in an RTA infra-red heating furnace. These glass-ceramic component solutions were allowed to gel, and the gels were dried at 130°C in air for the TGA-DTA measurements. Powders were prepared for XRD, SEM, and BET surface area measurements by calcining the dried gels. When a ramp rate higher than 20°C/min was employed, the sample was baked at 400-500°C to eliminate organic substance before firing. Baking and firing were performed in a clean room (Class 1000) and spin coating was performed on a clean bench (Class 100) to avoid dust and foreign particles.

When preparing a thin film on a metal foil capacitor, the solutions were applied by dip-coating technique on an Al foil (Nilaco, thickness (t) = 0.2mm), a

Colloidal Ceramic Processing

Ti foil (Nikkou Kinzoku, t = 0.054mm), and an 18-8 stainless steel foil (SUS) (Sumitomo Kinzoku, t = 0.052mm), followed by drying at 150-450°C on a heated plate. This procedure was repeated several times to obtain a desired thickness. Then the samples were put on a platinum plate and fired at 500-650°C at a heating and cooling rate of 600°C/min in an electric furnace.

Characterization

Gold was evaporated through a mask to form top electrodes (0.3 mm in diameter) for the electrical measurement. The yield (unshort-circuit percent) was calculated by the number of non-short-circuited electrodes out of about 40 top electrodes per sample. Dielectric constant and tanδ were measured at 0.1 V and 1 KHz unless otherwise specified by a precision LCR meter and an RF impedance analyzer. In case of metal foil capacitor, gold, nickel, and chromium were used as top electrodes (size:1x1.5, 1.5x1.8, or 3x4mm(for the high frequency impedance measurement)). The thickness of the film formed on the foils was assumed to be the same as that formed on a silicon wafer with the same dipping. As a result, the dielectric constant, the applied voltage, and the breakdown voltage are estimated values. The remnant polarization and the coercive field were measured by the Sawyer Tower circuit using a digitizing oscilloscope and a multifunction synthesizer.

The microstructure and the element analysis (EDX) of the thin film were performed by SEM, FE-SEM and STEM. The crystallization behavior of the powder was observed by XRD and TG-DTA. Glancing-angle XRD measurement was performed on fired thin films. The film thickness was measured by an etching technique. A portion of the thin film was masked by a methylmethacrylate polymer coating, followed by an etching with a hydrofluoric acid-hydrogen peroxide ($HF-H_2O_2$) solution, and the etched step height was measured by a surface profilometer at five different points and averaged.

Crystallinity is expressed by the relative (101) peak intensity to that of pure PT powder fired at 800°C. Semiquantitative PT content was calculated by the relative PT (101) peak area of samples fired at 500, 600, and 700°C for 30min ((101) peak area of pure $PbTiO3$ powder fired at 800°C = 100).

The designed volume fraction of the crystal phase was calculated assuming that every Ti atom was consumed along with an equimolar amount of PbO to form PT, and that the remaining PbO, B_2O_3 and SiO_2 formed the glass phase. The densities of the components are 7.96 kg/m^3 for PT, 7.55 kg/m^3 for PZT, 5.5 kg/m^3 for PB, 6.5 kg/m^3 for 2PbO•B_2O_3 and 6.0 kg/m^3 for PbO•SiO_2[19].

RESULTS AND DISCUSSION

Crystallization Behavior of Glass Ceramics

TGA-DTA measurements were conducted first with the dried gel and then again after calcining the gel at 600°C to eliminate the influence of the organic components. The TGA measurements indicate that most of the organics burned

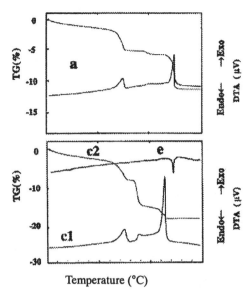

Fig.4 TG-DTA plots of precursor gels ("a" is a TG curve for PT, "c1" and "c2" are respective DTA and TG curves, and "e" is a DTA curve for calcined gels, of 0.8PT-0.2(2PB), respectively.

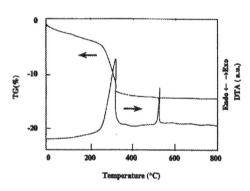

Fig.5 TG-DTA plots of precursor gels with designed composition of 0.7PT-0.3PbO·SiO₂.

off by 400°C. The endothermic peaks that likely reflect the melting of PbO-B₂O₃ components were observed at 577°C in the PB calcined powder and at 325°C in the 2PbO•B₂O₃ calcined powder. No endothermic peak was observed in the PT-PB calcined powder, suggesting an altered composition of the liquid phase. In the PT-2PbO-B₂O₃ glass-ceramics, the endothermic peak was observed at a temperature higher than that expected for the 2PbO•B₂O₃(Fig.4). This shift suggests that the PbO to B₂O₃ ratio was smaller in the amorphous phase of the glass-ceramics than the intended composition.

The higher PbO to B₂O₃ ratio of the glass-forming component yielded the lower crystallization temperature 484°C for 0.8PT-0.2PB and 458°C for 0.8PT-0.2(2PB), which were lower than for the pure PT gel (493°C). The perovskite phase of PT was observed in trace amounts at 450°C, becoming more prominent at temperatures above 500°C, until only this phase could be observed at 600°C using pure PT (Fig. 6). In the glass-ceramics, the pyrochlore phase predominated in the PT-PB system at 500°C, whereas the perovskite phase predominated even at 480°C in the PT-2PbO•B₂O₃ system (Fig. 7). The XRD patterns showed the tetragonal phase more prominently. Crystallization occurred at ~520°C, as shown by the sharp exothermic peak in the typical TG-DTA plots of the dried gel that had the designed composition of 0.7PT-0.3PbO·SiO₂ (Fig. 5). Dried gels

Colloidal Ceramic Processing

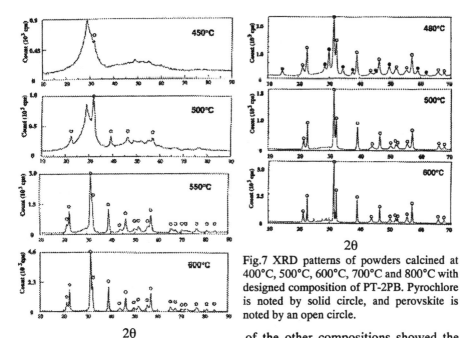

Fig.6 XRD patterns of PT calcined at (a) 450°C, (b) 500°C, (c) 550°C, and (d) 600°C. Perovskite is noted by open circle.

Fig.7 XRD patterns of powders calcined at 400°C, 500°C, 600°C, 700°C and 800°C with designed composition of PT-2PB. Pyrochlore is noted by solid circle, and perovskite is noted by an open circle.

of the other compositions showed the similar TG-DTA plots. The peak temperature became higher as the designed glass content increased (Table I).

In the case of silicate system, the glass-ceramic powder seemed to have a liquid phase, which is shown by an endothermic peak at ~700°C. As for the crystal phase, the pyrochlore phase appears at around 500-600°C as the first crystal phase, then a phase transition to the perovskite phase occurs at around 700°C (Fig. 7). The perovskite phase content decreased and the transition to the perovskite phase was retarded as the glass-forming phase increased (Table II). The perovskite phase was the major crystal phase at or above 700°C. There is no significant difference between the powder XRD patterns and the thin film XRD patterns.

The crystallization observed by DTA at ~520°C is probably caused by the pyrochlore-phase, considering the above-described XRD results. Because the pyrochlore phase seems to be thermodynamically slightly less stable than the perovskite phase from the fact that no corresponding peak is observed in the DTA plots, the transformation between two phases generally needs relatively high activation energy. The pyrochlore phase seems to result in less stress in the glass matrix than the perovskite phase when it crystallizes because the pyrochlore phase has a density closer to that of the lead silicate glass than does the perovskite phase. The pyrochlore phase precipitates at around 500°C at which the silicate glass still remains hard. The smaller volume change is apparently favorable. The pyrochlore phase remains as the major crystal phase until the glass softens. In the case of

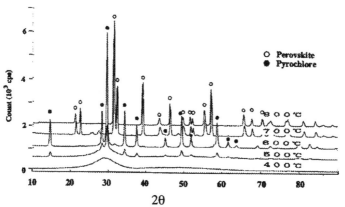

Fig.8 XRD patterns of powders calcined at 400°C, 500°C, 600°C, 700°C and 800°C with designed composition of 0.7PT-0.3PbO·SiO$_2$. Pyrochlore is noted by solid circle, and perovskite is noted by an open circle.

borate glass, because it softens at a temperature lower than 500°C, which is lower than the crystallization temperature, no volume change problem can occur. The perovskite phase will precipitate at first because it is more stable than the pyrochlore phase.

In case of PZT glass ceramic system, the examined compositions of the glass forming phase were 2PbO·B$_2$O$_3$, PbO·SiO$_2$, and 3PbO·SiO$_2$·B$_2$O$_3$. The crystallization behavior was very much similar to that of PT glass ceramic system.

Crystal Phase Content
The designed PT volume fractions and the obtained PT contents (fired at 700°C for 30 min) are summarized in Table I and II. At 500°C, no significant differences existed among the remaining samples at this temperature. As the calcination temperature increased, the PT contents also increased. The PT contents became roughly close to the designed PT volume fractions and seemed rather unaffected by the composition of the glass-forming component when fired at 700°C. This indicates that the chemical potential of Pb in the glass phase were close to that in the PT crystal phase in this experiment. As the calcination temperature increased, the PT contents also increased. The PT contents became more similar to the designed PT volume fractions when fired at 700°C.

Table I. Crystallization behavior of various PbO-TiO$_2$-B$_2$O$_3$ compositions.

PbO/ TiO$_2$/B$_2$O$_3$ (mol%) (Glass Phase)	Designed PTcontent (vol%)	PT content* (%)	TG-DTA peak (°C)	Crystallinity (I/I0)** (%)
5050/0 (None)	100	100	493	100
50/40/10 (PB)	74	99	484	100
50/27/23 (PB)	46	38	477	37
60/20/20 (2PB)	32	46	411	59
50/30/20 (PB)	52	40	470	46

*Semiquantatative PT content

Table II PbO-TiO$_2$-SiO$_2$ system.

Sample number	Designed composition	Designed PT content (vol %)	Measured PT content* (vol%) 500°C	600°C	700°C	DTA peak (°C)
2	0.9PT-0.1PS	87	7	9	78	516
3	0.8PT-0.2PS	76	0	3	25	521
4	0.7PT-0.3PS	65	0	2	46	525
5	0.6PT-0.4PS	55	0	2	13	529

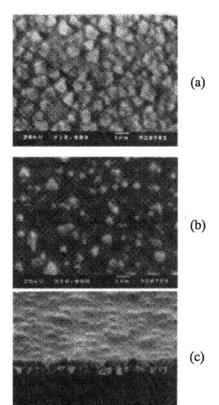

(a)

(b)

(c)

Fig.9 SEM micrographs of 0.54PT- 0.46(2PB) thin films fired at (a) 700°C*1min, (b) 800°C*1min, and (c) 800°C*1min (perspective view).

In the PZT glass ceramic system, the measured PZT volume fraction was close to the designed volume fraction and increased as the calcination temperature increased. The effect of borate and silicate were almost the same as observed in the PT system.

Microstructure

The SEM micrographs of fired thin films are shown in Fig. 9 for the borate glass system and in Fig. 10 for the silicate glass system. In the borate glass system, small crystals have developed in a glass matrix in every thin film that was fired at 700°C (Fig.9 a-c). These crystals seemed to be the perovskite type PT from the XRD results. The grain growth was observed when the thin films were fired at 500°C, then it became prominent at 600°C probably because of the liquid phase that enhanced the mass transfer. There was not a significant difference between the films fired at 700°C and 800°C.

In the silicate glass system, A defect-free, vitreous thin film was obtained when fired at temperatures >700°C, by the RTA process, with a designed glass phase of >10 mol% (Fig.10). When the firing temperature of the film was <600°C, films had a flat surface or a pore-grain structure. The

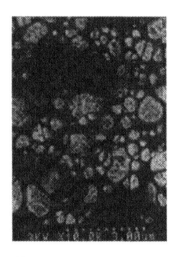

Fig.10 SEM micrographs of 0.7PT-0.3PS thin film fired at 700°C

higher the firing temperature, the larger the crystallite size became.

The crystallite size seemed to be larger as the glass content increases. Within the range of compositions tested, the glass-forming component of $2PbO \cdot B_2O_3$ showed the most enhanced crystal growth. Also, the larger the glass content was, the larger the crystallite size became probably due to the better mass transfer. Sometimes a film surface became so rough because of the enhanced grain growth that the yield of the film became almost zero.

In order to obtain a homogeneous thin film, it is important to consider the film thickness and the compatibility between the liquid phase of the film and the substrate surface. Because of the vitreous nature of $PbO-SiO_2$ glass, the yields of the silicate films were excellent in most of the composition. The drawbacks may be the suppression of the evolution of the ferroelectric (perovskite) phase and the resulting high firing temperature for the crystallization. This characteristic is just the opposite of the $PbO-B_2O_3$ system.[18,19]

In PZT glass-ceramic system, small crystals developed in a vitreous glass matrix in every thin film forming a contrast to the pure PZT thin film that contains a lot of pores. The crystals seemed to be of the perovskite type PZT considering the XRD results. Grain size in thin films increased as the firing temperature increased. The grain size became markedly large at 800°C probably because a liquid phase enhanced mass transfer. Grain morphology largely depended on the composition of the glass-forming component. The borosilicate glass system that was introduced to control the crystallization behavior better using B_2O_3 as a crystallization promoter and SiO_2 as a crystallization suppressor worked as intended.

Table III Electrical properties of $PbO-TiO_2-B_2O_3$ thin films.

$PbO/TiO_2/ B_2O_3$ (GLASS PHASE)	Film thickness (nm)	Firing temp. (°C)	Firing time (min)	Yield (%)	Dielec. const	Dielec loss
50/50/0 (NONE)	400	700	0.5	96	192	0.27
50/27/23 (PB)	520	700	1	88	181	0.053
60/20/20 (2PB)	270	700	1	92	122	0.061
50/30/20 (PB)	400	600	1	100	143	0.04

Colloidal Ceramic Processing

Electrical Property

When the thickness is small (100-200nm), the yield of a film was very low because of the reason discussed in the previous section. When a film was as thick as about 300nm, the yields were high in most of the films that contained high glass phase. In the case of the borate glass system, for example, the yield was 88%, ε=181, and tanδ=0.053(Table III). The dielectric constants obtained were extraordinarily high as glass-ceramics and were higher than the theoretical values. This is partly because the high ferroelectric phase content attained by this sol-gel glass-ceramic technique, and partly because the crystals in the glass-ceramics grew almost as

In the silicate glass system, the lowest yield was 91%. The dielectric constants were low when the films were fired at temperatures <600°C because the major crystal phase was the pyrochlore. The dielectric constant (ε) increased by a factor of ~2-3 to 200-300, and the dielectric loss (tanδ) was as low as 0.03-0.05 at temperatures >700°C, because the major crystal phase became the perovskite type.

The dielectric constants of the PZT glass-ceramic thin films on Pt/Ti/SiO2/Si substrates as a function of firing temperature are shown in Fig.11. The borate-containing system showed a high dielectric constant at a temperature higher than 650°C because of the high perovskite phase content. On the other hand, the silicate system showed a relatively low dielectric constant at a temperature lower than 750°C, which corresponded to the low perovskite phase content. The dielectric constants of the prepared glass-ceramics were generally much higher than expected.

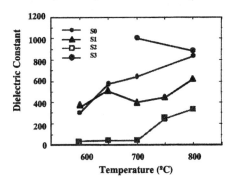

Fig.11 Dielectric constant of PZT glass-ceramic thin films as a function of perovskite content. (●)PZT, (▲) PZT-PbO-B$_2$O$_3$, (■)PZT-PbO-SiO$_2$, and (◗) PZT-PbO- B$_2$O$_3$-SiO$_2$)

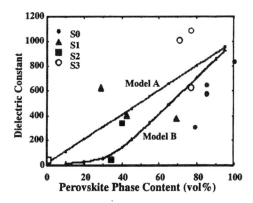

Fig.12 Dielectric constant of PZT glass-ceramic thin films as a function of perovskite content. (●)PZT, (▲)PZT-PbO-B$_2$O$_3$, (■)PZT-PbO-SiO$_2$, and (○) PZT-PbO- B$_2$O$_3$-SiO$_2$)

The dielectric constants are plotted against the perovskite phase volume fraction with the theoretical curves based on two different models; a parallel mixing model (model A) and another model in which the high dielectric constant spheres are randomly dispersed in a matrix (model B) (Fig.12). The dielectric constant of the ferroelectric phase of PZT was assumed to be 1000. Since the dielectric constant plots were scattered between and around the two theoretical curves, the reason was speculated as follows. When the crystallites are well developed and their diameters become close to the film thickness, the parallel mixing model would be appropriate. When the crystallites are small, a random sphere model would be better.

PZT Glass-Ceramic Thin Film Capacitor

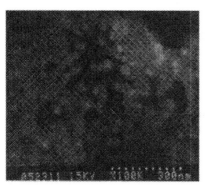

Fig.13 An SEM micrograph of the Al/PGC600

As an attempt of producing flexible thin film capacitors, dielectric layers were formed on metal foils. Gold, nickel, and chromium were sputtered through a mask to form the top electrodes on the PGC thin film on the Al foils (Al/PGC). The thin films were fired at 600°C for 1h (PGC600). The yields were generally good with all electrodes. The obtained small capacitance in Al/PGC600/Cr (substrate/dielectrics/top electrode) was probably due to the low-dielectric-constant reaction layer formed between the Cr electrode and the dielectric layer. Nickel seemed to be less active since the dielectric properties in Al/PGC600/Ni were similar to those with gold (which is inert).

To select suitable substrate materials, PGC thin films of various thicknesses were formed on SUS, Al, and Ti foils. Although the thermal expansion coefficients of the metal foils vary from 8.9 x 10^{-6} (Ti) to 24.8 x 10^{-6} (Al), no significant difference was observed in view of crack formation. Metal foils may be soft enough to relax the stress caused by the thermal expansion mismatch. However, the dielectric properties of the thin films strongly depended on the substrates.

In the case of SUS/PGC600/Au, insulation was excellent but the dielectric loss was too large. In the case of Ti/PGC/Au, the calculated dielectric constant (=500) were small compared with Si/SiO2/Ti/Pt (ε=1000, tanδ=3%). This is probably due to the low-dielectric-constant reaction layer formed between Ti and the glass-ceramics. This Ti capacitor showed relatively high dielectric constant partly because titania has a high dielectric constant (= 80) and partly because the reaction layer was thin as a result of the limited solubility of titania in glass.

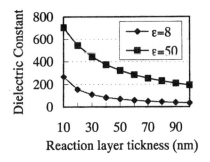

Fig. 14 Calculated dielectric constant of thin film capacitor on metal foils as a function of reaction layer thickness.

In the case of Al/PGC/Au, both the dielectric constant (=64) and the dielectric loss (= 0.1-0.9%) were very small. Because alumina forms glass more easily and dissolves in the glass more than titania, it is speculated that Al atom penetrated deep into the glass-ceramic layer and formed a compositionally graded, dense layer instead of the simple combination of the intermediate and the glass-ceramic layers, which then reduces the dielectric loss. The dielectric constant was low partly because the reacted glass-ceramic layer exists as mentioned above and partly because alumina has very low dielectric constant (= 8). The structure of the Al/PGC where very fine crystals are embedded in a dense glass matrix is shown in Fig. 13. The reaction layer thicknesses were estimated to be between 30-100 nm assuming the dielectric constant of the reaction layer as 8 for Al and 50 for Ti and the PGC layer as 1000(Fig. 14).

The precursor Ti/PGC samples were fired at 600, 650, 700, 750, and 800°C. The dielectric constants, the dielectric losses, the leak current and the impedance as functions of frequency and temperature were measured. The capacitance decreased as the firing temperature increased. An optimum firing temperature seems to lie between 600°C and 700°C.

Dielectric Properties

Gold electrode, the Ti and the Al foils were chosen for the further study after considering the experimental results on the top electrodes, the substrate material.

Impedance, capacitance, and tanδ of the Ti-foil capacitor as a function of frequency are examined. The capacitance was relatively high (10nF: ε*= 500), but tanδ was also high and they were stable only up to 300 kHz. There are moderate changes both in capacitance and tanδ of the Ti-foil capacitor as temperature increases. This type of capacitor may be used for the relatively large capacitance application.

The Al-foil capacitor can be used at

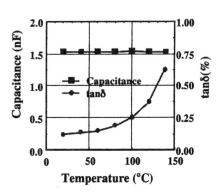

Fig.15 Capacitance and tanδ of Al/PGC/Au as a function of temperature. (PGC thickness=450nm, electrode area=2.7mm^2)

a very high frequency.   The capacitance was about 1.6nF ($\varepsilon^* = 64$) and stable in a wide frequency range (10^2-10^8Hz).   The dielectric loss was relatively low up to 1MHz but increased rapidly above 1 MHz.   The capacitance and tanδ of Al-foil capacitor showed excellent temperature stability (Fig.15).   Leak current was very low.   Three out of four capacitors showed very low leak current up to 5 V.   As a result, the Al-foil capacitor may be applied to a decoupling capacitor, or a built-in capacitor in a flexible print circuit board.

CONCLUSIONS
(1) Utilizing the sol-gel technique, achievement of crystal phase content at high levels hitherto unattainable by the conventional glass-ceramics technique was shown.   PT content was roughly close to the designed PT volume fraction and seemed rather unaffected by the composition of the glass-forming component when the glass-ceramics were fired at 700°C.
(2) PT and PZT were found to crystallize at temperatures blow 500°C. The structure in which crystals are embedded in a glass matrix was observed in thin films.   The PbO-B$_2$O$_3$ glass-forming component resulted in a low crystallization temperature and high crystalline content.   However, the grain growth enhancement sometimes resulted in poor insulation as a result of the surface roughness.   The PbO-SiO$_2$ glass forming component, which suppressed the crystal growth and necessitated a high crystallization temperature, turned out to have high insulation with a slightly lower dielectric constant than the PbO•B$_2$O$_3$ system.   The pyrochlore-phase was stabilized relative to the perovskite-phase until the liquid phase appeared in the system.
(3) Typical dielectric property of the glass-ceramic thin films are: the yield was 100%, ε=143, and tanδ=0.04 for the 0.6PT-0.4(PbO•B$_2$O$_3$) thin film, the yield was 100%, ε=198, and tanδ=0.044 for the 0.7PT-0.3PbO·SiO$_2$ thin film, the yield was 100%, ε=1000, and tanδ=0.035 for the composition of 0.835PZT-0.165(3PbO•SiO$_2$•B$_2$O$_3$) thin film.
(4) Thin film capacitors were successfully prepared by depositing 0.835PZT -0.165(3PbO•SiO$_2$ •B$_2$O$_3$) glass-ceramic thin films of 0.5 to 1 µm thickness on metal foils (0.05-0.2mm thickness) by a sol-gel process followed by firing at temperatures around 600ºC.   A relatively high capacitance (ε= 500) with a high dielectric loss (tanδ= 5%) capacitor was obtained with the Ti foil, and a relatively low capacitance (ε= 64) with a low dielectric loss (tanδ= 0.9%) capacitor was obtained with the Al foil.   Excellent ultra-high frequency properties (up to GHz) were shown in the case of Al foil capacitor.

REFERENCES
1) L. E. Sanchez, D. T. Dion, S. Y. Wu, and I. K. Naik, "Processing and Characterization of Sol-Gel Derived Very Thin Film Ferroelectric Capacitors", Ferroelectrics, **116** 1-17 (1991)

2) W. Park and M. Yamane, "Glass-Ceramics of the system Pb(Zr0.5Ti0.5)O3-GeO2", Glastech. Ber. **62**(5) 187-189 (1989)

3) B. Houng and M.J. Haun, "Lead Titanate and Lead Zirconate Titanate Piezoelectric Glass-Ceramics", Ferroelectrics, 154 (1-4), 1245-50 (1994)

4) Wu Mianxue and Zhu Peinan, "Piezoelectricity, pyroelectricity and ferroelectricity in glass ceramics based on PbTiO3", J. Non-Cryst. Solids, (84) 344-351 (1986)

5) D. G. Grossman, J. O. Isard, "Lead Titanate Glass-Ceramics", J. Amer. Ceram. Soc.,**52**(4) 230-231 (1968)

6) C. G. Bergeron and C. K. Russel, "Nucleation and Growth of Lead Titanate from a Glass", J. Am. Cer, Soc., **48**(3) 115-118 (1965)

7) T. Kokubo and M. Tashiro, "Fabrication of transparent PbTiO3 Glass-Ceramics", Bull. Inst. Chem. Res., Kyoto Univ., **54** (5) 301-306 (1976)

8) J. Marillet and D. Bourret, "Crystallization of lead titanate PbTiO3 in gel-derived solder glasses", J. Non-Cryst. Solids, **147&148** 266-270 (1992)

9) A. Herczog, "Microcrystalline BaTiO3 by Crystallization from Glass", J. Am. Ceram. Soc., **47** (3) 107-115 (1964)

10) T. Kokubo, C. Kung and M. Tashiro,"Preparation of Thin Films of BaTiO3 Glass-Ceramics and their Dielectric properties", Yougyou Kyokaisi **76** (4) 89-94 (1968)

11) Atit Bhargava et al., "Preparation of BaTiO3 Glass-Ceramics in the system Ba-Ti-B-O. I ,II", Materials Letters **7** (5,6) 185-196 (1988)

12) K. D. Budd, S. K. Dey and D. A. Payne, "Sol-Gel Processing of PbTiO3, PbZrO3, PZT, and PLZT Thin Films", Br. Ceram. Proc., 36 107-21 (1985),

13) C. Chen and D. F. Ryder, Jr., "Synthesis and Microstructure of Highly Oriented Lead Titanate Thin Films Prepared by a Sol-Gel Method", J. Am. Ceram. Soc., 72 [8] 1495-98 (1989)

14) S. L. Swartz, P. J. Melling and C. S. Grant, "Ferroelectric Thin Films by Sol-Gel Processing", Mat. Res. Soc. Symp. Proc., 152 227-32 (1989)

15) S. J. Milne and S. H. Pyke, "Modified Sol-Gel Process for the Production of Lead Titanate Films", J. Am. Ceram. Soc., 74 [6] 1407-10 (1991)

16) N. Tohge, S. Takahashi, and T. Minami, "Preparation of PbZrO3-PbTiO3 Thin Films by the Sol-Gel Process", J. Am. Ceram. Soc., 74 [1] 67-71 (1991)

17) S. S. Dana, K. F. Etzold, and J. Clabes, "Crystallization of Sol-Gel Derived Lead Zirconate Titanate Thin Films", J. Appl. Phys., 69 [8] 4398-403 (1991)

18) K. Saegusa, " PbTiO3-PbO-B2O3 Glass-Ceramics by Sol-Gel Process", J. Amer. Ceram. Soc., 79 [12] 3282-88 (1996)

19) K. Saegusa, "PbTiO3-PbO-SiO2 Glass-Ceramic Thin Films by a Sol-Gel Process", J. Am. Ceram. Soc., 80[10],2510-16 (1997)

20) The Society of Calorimetry and Thermal analysis, Japan, "*Thermodynamic Database MALT2*", Kagakugijutu-sya, Tokyo, (1993)

PREPARATION OF HIGHLY DISPERSED ULTRA-FINE BARIUM TITANATE POWDER BY USING ACRYLIC OLIGOMER WITH HIGH DENSITY OF HYDOROPHILIC GROUP

Yuichi Yonemochi, Yuichi Iida, Kenji Ogino, Hidehiro Kamiya
Graduate School of Bio-Applications and Systems Engineering, BASE
Tokyo University of Agriculture & Technology, Koganei, Tokyo 184-8588, Japan

Kenjiro Gomi, Kenji Tanaka
Murata Manufacturing Co., Ltd., Nagaokakyo, Kyoto 617-8555, Japan

ABSTRACT

In order to uniformly disperse ultra-fine $BaTiO_3$ particles with a stoichiometric composition and several 10nm in diameter to primary particles during the sol-gel synthesis process, an acrylic oligomer, which had a similar molecular weight and similar high density in hydrophilic carboxyl groups to the microbial-derived surfactant in our previous work, was prepared and added into synthesized solution. By using this oligomer, $BaTiO_3$ nano-particles with a high dispersion stability in suspension can be prepared.

INTRODUCTION

$BaTiO_3$, which is a typical material in capacitors, is widely used for multi-layer ceramic capacitors because of its high dielectric constant and ferroelectric properties.[1] For the miniaturization and high capacitance of multi-layer ceramic capacitors, many researchers are currently working on new processes for synthesizing ultra-fine $BaTiO_3$ particles from high-purity raw materials, for example, the sol-gel[2] and the hydrothermal methods.[3] Using these processes, it is rather easy to prepare ultra-fine $BaTiO_3$ powder that is less than 100nm in diameter and uniform in stoichiometric composition and that has a single phase. However, it is very difficult to keep the dispersion stability of these ultra-fine particles without the formation of hard and irregular aggregates in both aqueous and non-aqueous suspensions. In our previous paper[4], by the addition of microbial-derived surfactant during the synthesis process of particles, we succeeded in preparing BaTiO₃ nano-particles with a high dispersion stability in suspension. This surfactant has a relatively low molecular weight and high hydrophilic-group density and bending structure.

The present study focuses on the action mechanism of this surfactant. An artificial surfactant, which has a similar relatively low molecular weight and high density in hydrophilic carboxyl group without bending structure, was prepared and added into solution before synthesizing particles by sol-gel method. The action mechanisms of these surfactants were discussed based on the results of the present study and our previous work.

EXPERIMENTAL PROCEDURE
(1) Preparation and dispersion of BaTiO₃ powder

The barium hydroxide octahydrate and titanium tetraisopropoxide were used as starting materials for the ultra-fine BaTiO₃ powder and were prepared by Wako Pure Chemical Industries, Japan. The purity of barium hydroxide octahydrate and titanium tetraisopropoxide was 98% and 95%, respectively. Fig. 1 illustrates the preparation procedure of the ultra-fine BaTiO₃ with the surfactant. The 0.4M isopropyl alcohol solution of titanium tetraisopropoxide was prepared by diluting 11.4g of titanium tetraisopropoxide into isopropyl alcohol (Wako Pure Chemicals Industries, Japan) to 100ml. The 0.2M barium hydroxide aqueous solution was

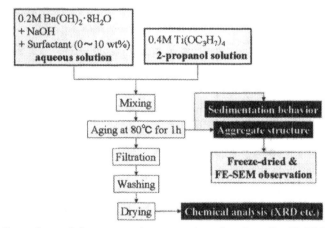

Fig. 1 A flow chart of the preparation procedure for ultra-fine BaTiO₃ particles with the addition of a surfactant during the synthesis process.

prepared by mixing 10.1g of barium hydroxide octahydrate, 156.7ml of distilled water, and 10.7ml of 12N sodium hydroxide solution (Wako Pure Chemicals Industries, Japan). The pH of the suspension that was made by the reaction of the barium hydroxide aqueous solution with the sodium hydroxide and the isopropyl alcohol solution of titanium tetraisopropoxide was about 13. Since the isoelectric point of $BaTiO_3$ is about 6 or 7, the prepared ultra-fine powder in suspension was charged negatively.

An acrylic oligomer, which is an artificial product shown in Fig. 2, was added to 0.2M barium hydroxide aqueous solution. The average degree of polymerization of this oligomer was about 10.5. This oligomer has a high density of hydrophilic group and without a bending structure. By using this surfactant and the same preparation process as in our previous work[4], $BaTiO_3$ nano-particles were prepared.

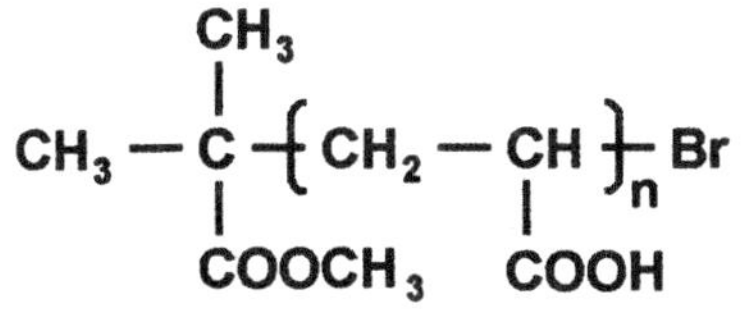

$$CH_3 - \underset{\underset{COOCH_3}{|}}{\overset{\overset{CH_3}{|}}{C}} - \left(CH_2 - \underset{\underset{COOH}{|}}{CH} \right)_n Br$$

Fig. 2 Estimated molecular structure of the artificial surfactant with a relatively low molecular weight and high hydrophilic carboxyl group density (n = 10.5).

To determine the optimum additive condition of this surfactant for the preparation of ultra-fine powder while maintaining the high dispersion stability and stoichiometry of the Ba/Ti ratio, the additive content of the surfactant to the synthesized $BaTiO_3$ particles was changed in the range of 0 to 10wt%.

The temperature of the barium hydroxide aqueous solution with sodium hydroxide was kept at 80°C with stirring. In order to prevent the generation of $BaCO_3$ by the reaction between barium hydroxide in solution and CO_2 in air, nitrogen gas was blown into the solution, and the CO_2 was purged. After the distribution of nitrogen gas was maintained for 10min with stirring in order to almost completely purge the CO_2, the isopropyl alcohol solution of titanium tetraisopropoxide was dropped into the barium hydroxide aqueous solution with sodium hydroxide and the surfactant, and the hydrolysis reaction of titanium tetraisopropoxide and the condensation reaction of $BaTiO_3$ were started. The amount of isopropyl alcohol solution of titanium tetraisopropoxide for the barium hydroxide aqueous solution was adjusted so that the mole ratio of Ba/Ti would be equal to 1.01. After aging by blowing nitrogen gas and mixing for 90 min (60 min with heating to maintain the temperature at 80°C and 30 min without heating to cool the suspension), the hydrolysis and condensation reaction of $BaTiO_3$ could finish. The final solid concentration in suspension was about 1vol%.

(2) Dispersion stability and aggregate structure characterization in suspension

The dispersion stability and aggregate structure of prepared particles in suspension were characterized by the following two methods. Firstly, the sedimentation behavior of synthesized particles in suspension was characterized. One hundred ml of synthesized suspension was transferred after aging for 90min in a graduated cylinder where it stood in a vibration-proof state. The time-dependent change of the dense-layer volume was measured during the sedimentation process.

Another method to characterize the aggregation of the synthesized powder was direct observation of the aggregate structure of the prepared particles in suspension by a process using field emission scanning electron microscopy, FE-SEM. In order to maintain the accurate aggregate structure of the particles in suspension, the part of the synthesized suspension after aging for 90min was transferred to a small glass container, and the circumference of the container was rapidly cooled using liquid nitrogen so that the suspension was frozen and solidified. After the suspension was completely solidified, a vacuum was formed inside the glass container. The microstructure of the aggregates of the freeze-dried samples of synthesized powder was observed using FE-SEM.

(3) Primary particle characterization

For characterization of the primary particle structure and chemical compound, the synthesized suspension after aging was filtered and washed with de-ionized water a few times to remove the sodium ions in the filtered cake. This filtered cake was dried at 100°C for 12h. The crystalline phases and the chemical composition of the dried powder sample were identified by X-ray diffraction (RAD-IIC, Rigaku, Tokyo, Japan) and X-ray fluorescence analysis (SMX10, Rigaku, Tokyo, Japan), respectively. The primary particle size was calculated from X-ray diffraction data by using the Scherrer's equation.

Colloidal Ceramic Processing

RESULTS

(1) Effects of the addition of surfactant on the primary particle structure of synthesized ultra-fine BaTiO₃ powder

Fig. 3 shows the XRD patterns of the synthesized ultra-fine BaTiO₃ powder and the effect of the additive content of the surfactant in the solution on the crystal phase. The Ba/Ti molar ratio of the synthesized ultra-fine powder with the sol-gel method was determined by X-ray fluorescence data in Table 1. The synthesized powder in suspension was washed after filtration to remove the sodium ions and then dried at 100℃. The content of the surfactant additive varied within a range of 0~10wt% for the synthesized ultra-fine BaTiO₃ particles. In our previous work, the excess addition (> 7wt%) of the microbial-derived surfactant inhibited the synthesis reaction of BaTiO₃ with the heterogeneous phase observation and the lower Ba/Ti molar ratio of the stoichiometric composition. However, in case of the artificial surfactant, the synthesis reaction has not inhibited in the range from 0 to 10wt% and the cubic perovskite phase of BaTiO₃ was observed and the Ba/Ti molar ratio was almost 1.

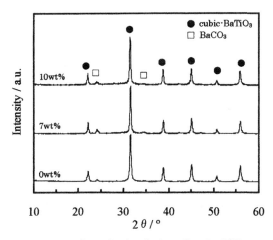

Fig. 3   XRD patterns of synthesized ultra-fine BaTiO₃ powder with or without a surfactant during the synthesis process.

Table 1 Chemical composition of a synthesized ultra-fine BaTiO₃ powder with or without the addition of a surfactant during the synthesis process

Additive content of surfactant [wt%]	0	7	10
Ba/Ti [-]	0.998	0.943	0.975

(2) Dispersion stability and aggregation behavior of the prepared BaTiO$_3$ powder in suspension

In order to discuss the dispersion stability and aggregation behavior of synthesized ultra-fine particles in suspension, the sedimentation behavior of each prepared suspension with different additive conditions of surfactant was measured and is shown in Fig. 4. For the suspension without the addition of a surfactant, the sedimentation of the particles in suspension started immediately after the solution was transferred to a graduated cylinder. The sedimentation of the ultra-fine BaTiO$_3$ particles was almost completed within 40min of starting the measurement.

On the contrary, the addition of a surfactant reduced the sedimentation rate of particles in suspension. The sedimentation rate of particles in suspension decreased with an increase in the content of the surfactant additive in solution. The volume ratio of the dense layer to the total suspension volume after 180min was increased from 32vol% (without the surfactant) to 86vol% (7wt% addition) and 91.5vol% (10wt% addition). The dispersion stability of the synthesized ultra-fine BaTiO$_3$ particles in suspension increased with an increase in the content of the surfactant additive.

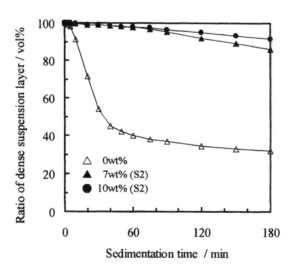

Fig. 4 Effect of additive content of surfactant on the sedimentation behavior for ultra-fine BaTiO$_3$ particles in prepared suspension.

To discuss the mechanism of dispersion stability by the addition of surfactant, Fig. 5 shows an FE-SEM magnification of the aggregate structure of synthesized primary particles in suspension, the prepared suspension after aging for 1h was freeze-dried. The mean primary particle diameter observed by FE-SEM was about 30~40nm, a clear difference in the primary particle diameter could not be recognized in an FE-SEM magnification. Fig. 5 (a) shows the aggregate diameter prepared in solution without a surfactant was distributed within the 10~50μm range. The detailed microstructure of these aggregates was observed with high-resolution FE-SEM as shown in Fig. 5 (b), the primary ultra-fine BaTiO₃

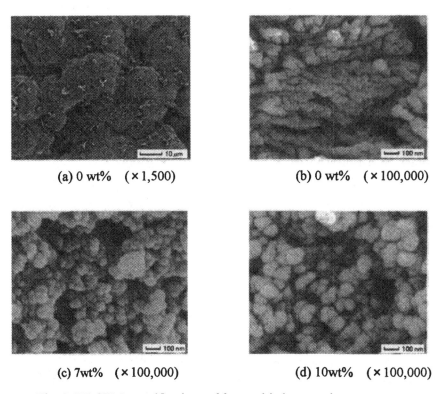

(a) 0 wt%　(×1,500)　　　　　　　　(b) 0 wt%　(×100,000)

(c) 7wt%　(×100,000)　　　　　　　(d) 10wt%　(×100,000)

Fig. 5　FE-SEM magnifications of freeze-dried suspensions.
The additive content of surfactant in synthesized solution was (a) 0wt% at 1,500 times, (b) 0wt% at 100,000 times, (c) 7wt% at 100,000 times and (d) 10wt% at 100,000 times magnifications.

particles formed a sparse and branch-shaped aggregate. These branch-shaped aggregates inhibited the dense packing of primary particles and caused an increase in the void in the inter-particle.

On the other hand, since the aggregate diameter was reduced by the addition of the artificial surfactant, the size of aggregates were between $100\sim200$nm in diameter and were observed at higher magnification (Fig. 5 (c), (d)).

Observations of the freeze-dried suspension confirmed that addition of an optimum amount of surfactant in solution reduced the size of the aggregates from several 10μm to $100\sim200$nm in diameter and changed the aggregates from sparse and branch-shaped to a closely packed structure. Since the addition of the surfactant caused a reduction in the size of the aggregates and a decrease in the sedimentation rate of the aggregates in suspension, suspension with high dispersion stability without sedimentation was obtained.

DISCUSSION

Based on the above results, the action mechanism of this surfactant during synthesized process was estimated. Because the synthesized reaction needs to keep a high pH value >13 in solution, the hydrophilic carboxyl groups of the surfactant are fully dissociated and charged negatively. Since the isoelectric point[5] of $BaTiO_3$ particles is about pH=$6\sim7$, the surface of the synthesized ultra-fine $BaTiO_3$ particles was also charged negatively in such high-pH conditions. Therefore, it is difficult for the negatively charged surfactants to electro-statically adsorb on the negatively charged surface of the $BaTiO_3$ particles after synthesizing the $BaTiO_3$ particles.

The artificial surfactant has a relatively low molecular weight and high density of hydrophilic carboxyl. Firstly, because the surfactant was mixed into barium hydroxide aqueous solution in a high pH condition, almost all COOH groups in this surfactant dissociated into COO^- in the water, COO^- groups in surfactant seemed to have formed the complex with the Ba^{2+} ion in the barium hydroxide aqueous solution with high pH conditions. Because of the nucleation by the reaction with Ba^{2+} ion in complex and diffused $Ti(OH)_6^{2-}$ ions and growth of ultra-fine $BaTiO_3$ particles produced in this nano-scaled water-pool space in solution, the aggregation of primary particles was suppressed, and the suspension stability was increased by the addition of the artificial surfactant.

Colloidal Ceramic Processing

CONCLUSION

By using an artificial surfactant, which had a similar molecular structure to the microbial-derived surfactant in our previous work, ultra-fine BaTiO₃ particles with high dispersion stability in suspension and uniform stoichiometric composition were prepared. The addition of the high additive content, up to 10wt% did not inhibit the synthesis reaction of stoichiometric BaTiO₃ particles.

ACKNOWLEDGEMENT

This study was supported by a Grant-in-Aid for Scientific Research (B) from the Japanese Ministry of Education, Science, Sports and Technology, and the Murata Science Promotion Foundation, and the Structurization of Material Technology Project in the Nano-Technology Program by METI Japan.

REFERENCE

1. L. E. Cross, "Dielectric, Piezoelectric and Ferroelectric Components," *Am. Ceram. Soc. Bull.*, **63**, 586-90(1984)

2. K. W. Kirby, "Alkoxide Synthesis Techniques For BaTiO₃," *Mater. Res. Bull.*, 23, 881-90 (1988)

3. G. Guldner, T. Heinrich and D. Sporn, "Hydrothermal Synthesis of Ferroelectric Powders," *Euro. Ceram.*, **3**, 1873-77 (1991)

4. H. Kamiya, K. Gomi, Y. Iida, K. Tanaka, T. Yoshiyasu and T. Kakiuchi, Preparation of highly dispersed ultra-fine barium titanate powder by using microbial-derived surfactant", *J.Am.Ceram.Soc.*, accepted.

5. U. Paik and V. A. Hackley, "Influence of Solid Concentration on the Isoelectric Point of Aqueous Barium Titanate," *J.Am.Ceram.Soc.*, **83**(10), 2381-84 (2000)

THE THERMAL STABILITY AND STRUCTURAL PROPERTIES EVOLUTION OF CURED AND NON-CURED ZrO_2 AND ZrO_2-SiO_2 POWDERS

Qiang Zhao, Wan Y. Shih, and Wei-Heng Shih
Materials Science and Engineering
Drexel University
Philadelphia, PA 19104

ABSTRACT

The curing effect on ZrO_2 powders was studied using Teflon flask to avoid the Si dissolved from glass containers. It was found that curing ZrO_2 powders at 75°C for 5 days generates larger pore volume and pore size. The cured ZrO_2 also has 30~40 m^2/g higher specific surface area than non-cured ZrO_2 before extensive heat treatment. As heat treatment time increases ZrO_2 powders begin to sinter. At the same time the pore volume decreases and the pore size increases and the initial surface area difference between the cured and non-cured ZrO_2 diminishes with time. On the other hand, for Si-doped ZrO_2 powders, the initial surface area difference between cured and non-cured samples is preserved even after hours of calcinations at elevated temperatures because Si effectively inhibits sintering and stabilizes the textural structures of ZrO_2 powders.

INTRODUCTION

Fine zirconia powder is an important raw material for structural components and catalysts. Zirconia has superior toughness due to the transformation toughening mechanism. Zirconia is also an important catalyst promoter in automobile exhaust catalytic system due to its high oxygen transport. For catalytic applications, thermal stability and high surface area at elevated temperature are required.

There have been several studies in promoting the thermal stability of ZrO_2. In the synthesis of ZrO_2 powders by wet chemistry methods, Stichert and Schuth [1] showed that using lower concentration precursor generates smaller crystallite size and higher surface area powders. Chuah et al. [2,3,4] showed that curing enhanced the surface area of ZrO_2 at elevated temperatures. Curing of ZrO_2 in the mother solution at high temperature was speculated to enhance the hydroxylation bonding between the ZrO_2 particles resulting in smaller weight loss during heat treatment and the stronger structure maintains larger surface area at elevated temperatures. Using SAXS Zeng et al. [5] show cured hydrous zirconia precipitates after drying had more branchy particle structure (low fractal dimension), smoother surface and larger surface area compared with non-cured samples. They believe it is the dissolution and reprecipitation of hydrous zirconia powders during curing that causes these differences.

However, it was shown recently that curing at high pH conditions might lead to the dissolution of Si from the container (both stainless steel and glass type)[6] or any other part that was contact with the mother liquor and contained Si[7]. It is well known that Si doping inhibits ZrO_2 grain growth, delays the crystallization, and enhances the surface area of ZrO_2 powders at high temperatures [8,9]. In the literature [2-5] the pH of the curing suspensions were >9 and the containers used for curing were either not specified or made of Pyrex glass. It is likely the results reported by these authors were the combined effects of curing and Si doping. If the high surface area and thermal stability of cured zirconia are attributed to the Si contamination according to reference [6] and [7], the question whether the pore volume and pore size change for cured zirconia observed in [4] and [5] is also due to Si effect arises. We performed a systematic study on the effects of curing and Si on ZrO_2 powders using Teflon flasks [10], which eliminated the presence of Si in the system from dissolution. Si will be added through hydrolysis of TEOS solution in a control manner when it is needed. The conclusions from our study were cured zirconia powders have larger pore size and pore volume compared

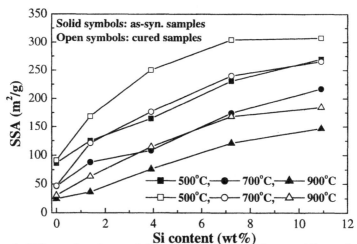

Fig. 1. Effect of curing and Si concentration on the specific surface area of ZrO_2 powders. All samples are heat treated for 4 hours at indicated temperatures before measurement.

with non-cured zirconia whether Si was present or not. Without Si doping, cure zirconia cannot maintain its higher surface area (difference <4%) than non-cure zirconia after 4 hours calcinations (see figure 1, zero Si concentration points), which was consistent with the speculation derived from [6] and [7]. However, when Si was added to ZrO_2, cured samples always have higher surface area than non-cured samples (~20 to 30% difference) (see figure 1). In this paper, we investigated the reasons why Si doping increases the surface area of cured ZrO_2 more than the non-cured ZrO_2 by tracing the textural properties evolution of pure ZrO_2 and Si-doped ZrO_2 at different heat treatment time.

EXPERIMENTAL PROCEDURE

The sample preparation was the same as the previous paper [10]. 50 ml of 0.33 M $ZrO(NO_3)_2$ solution was added drop by drop to a Teflon flask containing 200 ml of 5.02 N NH_4OH with rigorous magnetic stirring. Zirconium hydroxide was precipitated due to the high pH condition. After the $ZrO(NO_3)_2$ solution was added, the suspension was kept stirring for half hour. The precipitates were centrifuged and dried at 80°C in air overnight. The obtained hydrous zirconia powders are labeled the as-synthesized ZrO_2. Curing was performed after precipitation but before centrifugation by heating the precipitates in the mother liquid to 75°C and holding for 5 days. The addition of Si was performed by adding 0.42 g of TEOS:2-propanol:H_2O (1:3:1 molar ratio) solution dropwise to the as-synthesized or the cured zirconium hydroxide suspensions before centrifugation and stirring for another half hour. The final Si content in ZrO_2 is 1.38 wt%. Powders were collected by centrifugation, followed by drying at 80°C in air similar to the as-synthesized ZrO_2. The centrifuged supernatants of the Si doped samples were tested by ICP to make sure there was no Zr or Si loss during processing. As a result, the final compositions of the powders are consistent with the nominal values and the comparison between the cured and the as-synthesized samples is based on the same Si content. Calcination of samples was conducted in air at 500°C for various times by a heating rate of 5°C/min. The 0 and 12 minutes heat treated samples were air quenched to room temperature by pulling the sample out of the furnace directly and the other calcined samples were cooled to room temperature in the furnace. The powder X-ray diffraction (XRD) patterns of the samples were obtained by a Siemens D-500 X-ray diffractometer using Cu Kα radiation (1.5416Å). The crystallite size were calculated by the (111) peak of tetragonal phase, according to the Scherer formula,

$$\varepsilon = \frac{0.9\lambda}{\beta \cos \theta} \tag{1}$$

where ε is the apparent crystallite size, λ is the wavelength of radiation, $\beta=(B^2-b^2)^{1/2}$ is the true broadening of FWHM (in radians), B is the measured FWHM of sample, b is the broadening of FWHM due to the instrument, and θ is the Bragg angle. Nitrogen-gas physisorption measurements were performed with a Quantachrome Nova 2200 Surface Area Analyzer at -196°C. To determine the pore size distribution, desorption isotherm was analyzed with the BJH method, while specific surface areas were determined using the BET method (P/P_0=0.05-0.3).

RESULTS

The surface areas of ZrO_2 and Si-doped ZrO_2 as a function of heat treatment time are shown in Fig. 2(a) and (b). The surface areas decrease as the heat treatment time increases. For pure ZrO_2 there is a ~38 m^2/g surface area

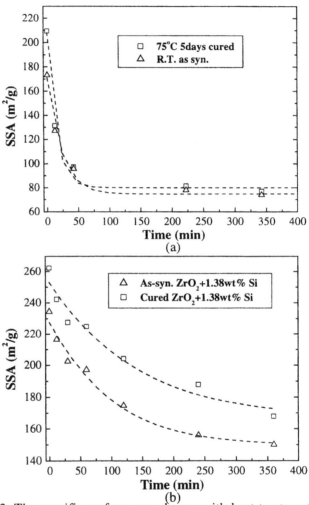

Fig. 2. The specific surface area change with heat treatment time of pure ZrO_2 (a) and Si doped ZrO_2 (b) at 500°C. The dash lines in the plots are just for guiding the eye.

difference between the cured sample and the as-synthesized sample initially. The surface area difference diminishes very quickly (within 12 minutes) as the heat treatment time increases. After 6 hours of heat treatment, the final surface area is only 40% of the initial surface area. On the other hand, for the Si-doped samples, there is a ~30 m²/g surface area difference between the cured and the as-synthesized samples at all times. Considering the very low concentration of Si

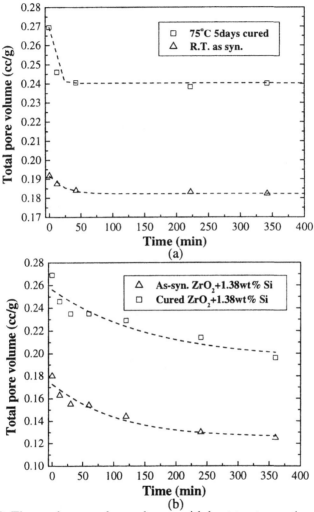

Fig. 3. The total pore volume change with heat treatment time of pure ZrO₂ (a) and Si doped ZrO₂ (b) at 500°C. The dash lines in the plots are just for guiding the eye.

doping and the similar value with the initial surface area difference of pure ZrO₂ 30 vs. 38 m²/g), the surface area difference of 30 m²/g is expected to be due to

ZrO_2 and not SiO_2. And after about 6 hours heat treatment, the final surface area of Si-doped samples is approximately 65% of the original surface area, indicating Si retards the sintering of ZrO_2 compared with pure ZrO_2 powders.

The evolution of pore volume with heat treatment time is shown in Fig. 3 (a)

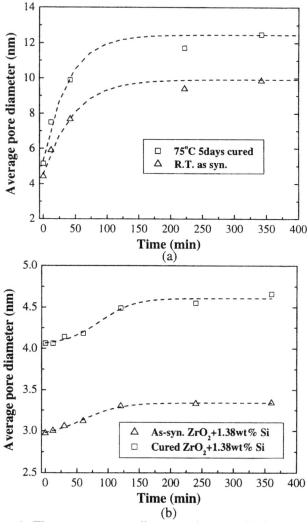

Fig. 4. The average pore diameter change with heat treatment time of pure ZrO_2 (a) and Si doped ZrO_2 (b) at 500°C. The dash lines in the plots are just for guiding the eye.

and (b). The pore volume decreases as the heat treatment time increases due to the sintering of powders. The cured samples have higher pore volume throughout the heat treatment time for both Si-doped and undoped ZrO_2. The average pore size increases with heat treatment time (Fig. 4 (a) and (b)). This is due to the fact that smaller pores disappear first. The cured samples have larger pore size than the as-synthesized samples, consistent with the results of pore volume. With Si doping the pore size of ZrO_2 powders increases much slower than without Si doping. Table I and II shows the phase evolution of pure and Si doped ZrO_2. Pure as synthesized ZrO_2 already crystallizes to tetragonal phase when temperature reaches 500°C and the tetragonal phase transforms to the monoclinic phase as time increases. Pure cured ZrO_2 is amorphous initially and becomes tetragonal and monoclinic phases as heat treatment time increases. The Si doping delays the crystallization of ZrO_2 powder and stabilizes the tetragonal phase of ZrO_2, so there is no tetragonal-monoclinic phase transformation for Si-doped ZrO_2. It is commonly accepted that nanosize tetragonal ZrO_2 can exist at room temperature because of its small size. When the particle size exceeds a critical value, the tetragonal-monoclinic phase transformation will occur. So combining the results from Fig. 2, Fig. 4 and table I, II, it is clear that Si doping inhibits ZrO_2 sintering and grain growth, and the initial textural structure of ZrO_2 powders can be preserved better compared with that without Si doping.

Table I. The phase evolution of pure ZrO_2 at 500°C heat treatment

Time (min)	0	12	42	222	342
R.T. as syn.	T(7.7)	T+M	M+T(m)	M+T(s)	M+T(s)
75°C 5days cured	A	T+M	M+T(s)	M+T(s)	M+T(s)

Table II. The phase evolution of Si doped ZrO_2 at 500°C heat treatment

Time (min)	0	12	30	60	120	240	360
R.T. as syn.	A	A	A	A	T (9.4)	T (10.8)	T (10.4)
75°C 5days cured	A	A	A	A	T (6.9)	T (8.9)	T (9.5)

T: tetragonal ZrO_2, M: monoclinic ZrO_2, A: amorphous, m: middle amount, s: small amount, and the numbers in the parentheses are the crystallite size in nm.

As discussed in the previous papers [2-5, 10], curing process involves the condensation of non-bridged hydroxyl groups, particle dissolution and reprecipitation, and results in a stronger structure that can sustain the capillary force during drying. After drying or low temperature heat treatment, the cured samples will have more open structure and more accessible surface area than the non-cured samples. Indeed, in this study we found the cured samples always had larger pore volume and pore size than the as-synthesized samples. The initial surface area of the cured sample was actually higher than the as-synthesized sample. However, the initial structure couldn't withstand sintering, so the surface area difference diminished quickly during heat treatment. Only when Si was present in the ZrO_2 powders, which inhibited sintering and stabilized ZrO_2, the surface area difference between the cured and the as-synthesized samples can be maintained after high temperature and long time heat treatment (Fig. 1).

DISCUSSION

Using the surface area reduction during isothermal sintering, the sintering mechanism was deduced from the study of sintering kinetics [11,12]. One of the basic formulas used is:

$$\left(\frac{\Delta S}{S_0}\right)^N = Bt \tag{2}$$

where ΔS is the change in surface area from the initial value S_0, B is a function of powder properties and temperature, t is isothermal sintering time, and the exponent N characterizes the dominant sintering mechanism according to the following table.

Table III. Relation between sintering mechanism and exponent N

Sintering Mechanism	Approximate value of N
Viscous Flow or Plastic Flow	1.1
Evaporation-Condensation	1.5
Volume Diffusion	2.7
Grain Boundary Diffusion	3.3
Surface Diffusion	3.5

The plot of log ($\Delta S/S_0$) vs. log (t) for Si-doped ZrO_2 is shown in Fig. 5. The N values for the cured and the non-cured samples, which are the inverse of the

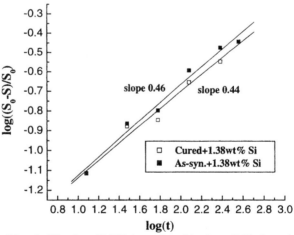

Fig. 5. The log ($\Delta S/S_0$) vs. log (t) plot of Si-doped ZrO_2 powders heat-treated at 500°C

Colloidal Ceramic Processing

slopes, are 2.3 and 2.2 respectively. According to the Table III, the dominant sintering mechanism would be volume diffusion for these two samples. However, we must be careful to draw such conclusion. The model used to derive equation (2) assumes spherical solid particles during the initial stage sintering. The powders we made were porous aggregates of small particles of irregular shape. And according to Table I and II, a series of phase transformations is also involved during heat treatment. Moreover the substantial surface area reduction (>35%) during heat treatment suggests that the sintering behavior is beyond the initial stage. So the power-law fit may not give the true mechanism. This can be seen more clearly from the log (ΔS/S$_0$) vs. log (t) plot for pure ZrO$_2$ (Fig. 6). The data cannot even be fitted into a straight line because pure ZrO$_2$ shows more significant surface area reduction (>60%) and phase transformation. It is likely that the sintering phenomena involve porous aggregates rather than solid particles.

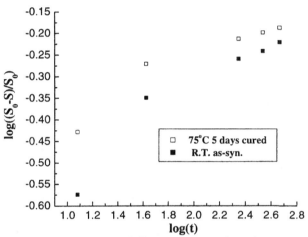

Fig. 6. The log (ΔS/S$_0$) vs. log (t) plot of pure ZrO$_2$ powders heat treated at 500°C

CONCLUSION

Curing ZrO$_2$ powders at 75°C for 5 days generates larger pore volume and pore size. The cured ZrO$_2$ also has 30~40 m^2/g higher specific surface area than non-cured ZrO$_2$ before going through extensive heat treatment at 500°C. As heat treatment time increases ZrO$_2$ powders sinter together. The pore volume decreases, pore size increases and the initial surface area difference between the cured and the non-cured ZrO$_2$ diminishes quickly. However, for Si doped ZrO$_2$ powders the initial surface area difference between the cured and the non-cured samples can be preserved even after hours of calcinations at elevated temperatures because Si effectively inhibits sintering and stabilize ZrO$_2$ textural structures.

REFERENCES

[1] W. Stichert and F. Schuth, "Influence of Crystallite Size on the Properties of Zirconia", *Chem. Mater.*, **10**, 2020-2026 (1998).

[2] G. K. Chuah, S. Jaenicke, S. A. Cheong, and K. S. Chan, "The Influence of Preparation Conditions on the Surface Area of Zirconia", *Appl. Catal. A*, **145**, 267-284 (1996).

[3] G. K. Chuah and S. Jaenicke, "The Preparation of High Surface Area Zirconia-Influence of Precipitating Agent and Digestion", *Appl. Catal. A*, **163**, 261-273 (1997).

[4] G. K. Chuah, S. Jaenicke, and B. K. Pong, "The Preparation of High-Surface-Area Zirconia", *J. Catal.*, **175**, 80-92 (1998).

[5] Y. Zeng and Y. Zhao, "An investigation of Submicrostructural Evolution of Freshly Precipitated ZrO_2 in Digestion for Ultrafine Sulfated ZrO_2 Catalysts by SAXS Technique", *J. Colloid. & interf. Sci.*, **247**, 100-106 (2002).

[6] H.-L. Chang, P. Shady, and W.-H. Shih, "The Effects of Containers of Precursors on the Properties of Zirconia Powders," accepted by *Microporous and Mesoporous Materials*.

[7] S. Sato, R. Takahashi, T. Sodesawa, S. Tanaka, K. Oguma, and K. Ogura "High-Surface-Area SiO_2–ZrO_2 Prepared by Depositing Silica on Zirconia in Aqueous Ammonia Solution", *J. Catal.*, **196**, 190–194 (2000).

[8] S. Soled and G. B. McVicker, "Acidity of Silica-Substituted Zirconia", *Catal Today*, **14**, 189-194 (1992).

[9] R. A. Shalliker, G. K. Douglas, L. Rintoul, and S. C. Russell, "The Analysis of Zirconia-Silica Composites using Differential Thermal Analysis, Fourie Transform-Raman Spectroscopy and X-ray Scattering/Scanning Electron Microscopy", *Powder Tech.*, **98**, 109-112 (1998).

[10] Q. Zhao and W.-H. Shih, H.-L. Chang and P. Andersen, "The Effect of Curing and the Thermal Stability of Si-doped ZrO_2 Powders", submitted to *Applied Catalysis A: General*.

[11] R. M. German and Z. A. Munir, "A Kinetic Model for the Reduction in Surface Area During Initial Stage Sintering"; *Sintering and Catalysis*, pp.249-257, edited by G. C. Kuczynski, Plenum Press, New York 1975.

[12] Randall M.German, "Solid-State Sintering Fundamentals"; pp.98-103 in *Sintering theory and practice*, Wiley, New York, 1996.

MICROPOROUS SILICA MODIFIED WITH ALUMINA AS CO_2/N_2 SEPARATORS

T. Patil, Q. Zhao, W. Y. Shih, and W.-H. Shih
Department of Materials Engineering, Drexel University, Philadelphia, Pennsylvania 19104-2875

R. Mutharasan
Department of Chemical Engineering, Drexel University, Philadelphia, Pennsylvania 19104

ABSTRACT

To reduce global warming, carbon dioxide needs to be removed from flue gases prior to release into the atmosphere. Since molecular size of CO_2 and N_2 are similar, the separation by pore size is not a viable approach. Differences in adsorption behaviors of CO_2 and N_2 have been proposed as a possible means for separation. In this paper, we investigated the synthesis of microporous aluminosilicates which have preferential adsorption of CO_2 over N_2. It was found that microporous silica modified with alumina by a sol gel method exhibits adsorption of CO_2 from its equimolar mixture with N_2 in gas chromatography (GC). The extent of adsorption is found to increase with decreasing temperature. Results indicate that adsorption with aluminosilicates occurred at a lower temperature than alumina alone. Silica modified with equal percentage of alumina was found to have high specific surface area. The adsorption isotherm reveals the presence of micropores as well as mesopores. The addition of a basic oxide BaO to the aluminosilicate matrix enhanced carbon dioxide adsorption, thus improving separation of CO_2. Increasing the percentage of the BaO increased the separation.

INTRODUCTION

The prospect of global climate change is a matter of public concern. The concentration of man-made CO_2 in the atmosphere has been increasing since the start of industrialization in the 19th century, and more rapidly during the second half of the 20th century. Even though there is uncertainty about the exact relationship between increasing CO_2 concentration and rising global temperatures, the connection is accepted by most people. And hence there is a growing concern for the removal of CO_2. To remove CO_2 effectively, it is necessary to separate CO_2 from N_2, the most abundant gas in the atmosphere. Over the past few years, the applications of gas separation membrane technology have gradually been adopted for CO_2 removal. Membranes made of different materials have been attempted. Polymeric membranes exhibit separation capability, but only at temperatures below 150°C. Ceramics have good thermal stability and have great potential for gas separation at high temperatures.

This study focuses on the development of membrane material that can separate CO_2 and N_2 at elevated temperatures. The membrane constructed with the material is expected to selectively pass CO_2 instead of N_2. Normal gas diffusion through a membrane includes several mechanisms such as molecular sieving and Knudsen diffusion.[1,2,3,4] The separation between CO_2 and N_2 cannot be based on molecular sieving because the kinetic diameters of CO_2 and N_2 molecules are similar (3.6Å and 4.0Å respectively). Knudsen diffusion selectivity bears an inverse proportionality with the square root of the molecular weights of CO_2 and N_2. Because the molecular weights of the two gases are not very different, the separation selectivity is low. We propose that surface selective adsorption along with molecular sieving could separate CO_2 and N_2 effectively. It is proposed that microporous silica will selectively sieve and adsorb CO molecules that will eventually diffuse through the pore due to a concentration gradient across the pores.

EXPERIMENTAL PROCEDURE AND RESULTS
Synthesis of Microporous Silica

Silica sols were first prepared by acid-catalyzed hydrolysis and condensation of tetraethylorthosilicate (TEOS- $(Si\,(OEt)_4)$) in ethyl alcohol.[5] 21 ml TEOS and 21 ml ethanol (molar ratio 1:3.8) were measured in a flask using a measuring cylinder. 10.8 ml distilled water and 8.11 ml nitric acid (98% HNO_3) (molar ratio 75:1) were mixed together in a cylinder. The (H_2O + HNO_3) mixture was slowly added to the (TEOS + ethanol)

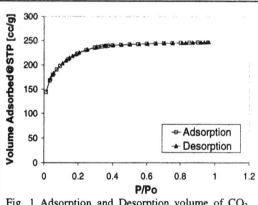

Fig. 1 Adsorption and Desorption volume of CO_2 adsorbed as a function of pressure.

solution kept in an ice bath at the rate of one drop every 5 seconds with vigorous stirring. The reaction mixture was then refluxed at 80°C for 3 hours. The refluxed mixture was allowed to cool down and dry in a Petri dish at room temperature obtain the silica. The silica sol gradually turned into a gel. The powder obtained after continuous drying were ground manually in a pestle and mortar and the heat treated at 400°C for 3 hours. Both the heating and cooling rate were kept at 0.5°C/min.

Surface Area Analysis

The microporous silica powders were analyzed by a BET surface-area analyzer (Quantachrom NOVA 2200) using N_2 physical adsorption at 77K, after outgasing at 300°C for 24 hours. The micropore surface area and approximate pore size were calculated by the t-method using Halsey equation. The Langmuir-type isotherm, given in Fig.1, indicates the existence of micropores. The t-plot of the same data shown in Fig. 2 gives a micropore surface area of 813 m^2/g and an approximate pore size of 10.5 Å.

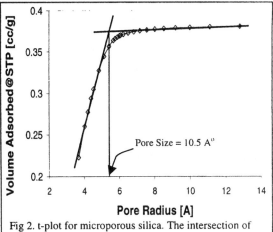

Fig 2. t-plot for microporous silica. The intersection of the slopes of the two curves gives the approximate pore diameter.

Gas Chromatography

The Gas Chromatography (GC) technique was used for characterizing adsorption of CO_2 on the microporous silica. The GC technique measures the differences in partitioning behavior of a gas mixture between a mobile phase and a stationary phase, by measuring retention or elution time. A column holds the stationary phase and the mobile phase carries the sample through it. Sample components that partition strongly into the stationary phase spend a greater amount of time in the column relative to non-adsorbing species, and are thus separated from other components resulting in separation. As the components elute from the column they can be quantified by a detector and/or collected for further analysis. We used argon as the mobile phase and CO_2/N_2 gas mixture as the stationary phase. The CO_2/N_2 gas mixture was injected using a four-way valve in order to control the amount of gas flowing into the column. GC was done on microporous silica at 50°C with a flow rate of 1 ml of CO_2/N_2 50/50 gas mixture per 36 seconds.

Figure 3 shows the GC separation for CO_2/N_2 gas mixture and pure N_2. Both chromatograms show only one peak with similar area under the peak indicating that the CO_2 is not retained by silica and elution from the column occurs at the

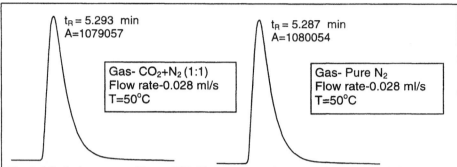

Fig. 3 GC of microporous silica at 50°C. The area under the peak with pure N_2 is similar to that with (1:1) CO_2-N_2 mixture suggesting no retention of CO_2. Also the retention times for both the peaks are the same suggesting the gas mixture elutes at the same time as pure N_2.

same time as the N_2 gas. Silica and CO_2 are both known to be acidic in nature and, thus it is hypothesized that silica has a weak affinity towards CO_2.

Impregnation of BaO

In order to improve the affinity of silica towards CO_2, we impregnate silica with BaO, which is a basic metal oxide.[8] 6.3% mole $Ba(OH)_2 \cdot H_2O$ was dissolved in distilled water and the solution was poured into 2g of microporous silica in a small tube and stirred for about half an hour with a spatula. The mixture was oven dried and then heat-treated at 400°C for 3 hrs at 0.5 °C/min. GC of the BaO impregnated silica was done at the same temperature (50°C) and flow rate as that for pure silica. It was observed that the GC plot showed only one peak. The same

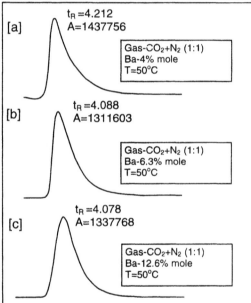

Fig. 4 GC of silica with different BaO%. [a], [b] and [c] have approximately the same area under the curve. Also the retention times are approximately the same. This suggests that there is no adsorption of CO_2 even in the presence of BaO.

procedure was repeated for three different percentages of BaO viz. 4%, 6.3% ar 12% in order to verify the effect of BaO on the adsorption of CO_2 on silica. Fig.

shows the GC plots for these three concentrations of BaO. It is concluded from Fig.4 that BaO addition to silica did not help the adsorption of CO_2.

Effect of Temperature

The GC results were also obtained at different temperatures to characterize the effect of temperature on the adsorption. Figure 5 shows the GC results for silica impregnated with 6.3% BaO at 25°C, 100°C, 150°C, 200°C. Since the area under the peaks is approximately same and also the retention times being similar for the plots at the four temperatures, it is clear that both CO_2 and N_2 are eluting simultaneously. These GC plots suggest that silica with BaO cannot bring about the separation of CO_2 and N_2 within the temperature range explored.

Synthesis of 1:1 Aluminosilicate Powders

In our earlier work (unpublished), it was found that alumina could adsorb CO_2 gas. With this background, a binary sol of silica and alumina was prepared and tested for any adsorption of CO_2. The binary sol was prepared using TEOS and aluminum-sec-butoxide as the alkoxide precursor. 10.5 ml TEOS and 8 ml of ethanol were mixed in a measuring flask. 1.52 ml H_2O and 8.11 ml 98 % HNO_3 were mixed in another flask. The acidic mixture was added to the TEOS and ethanol mixture dropwise at an approximate rate of a drop per 10 seconds with continuous stirring. After an hour 227 ml of ethanol was added to

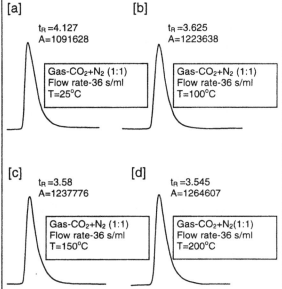

Fig. 5 GC of SiO_2+ 6.3% BaO at different temperatures. [a], [b], [c] and [d] suggest no adsorption of CO_2 even with the variation of operating temperature. The area under the peaks and retention times being the same suggests that temperature change between 25°C and 200°C does not affect the adsorption of CO_2.

2.35 ml of Al-sec-butoxide and stirred for 30 minutes. The acidic TEOS mixture was then added dropwise to the Al-sec-butoxide mixture again at a rate of a drop per 10 seconds with vigorous stirring. The sol was allowed to cool and dry in a petri dish to room temperature. The powder obtained after drying were ground manually in a pestle and mortar and then heat treated to 400°C for 3 hours. Both the heating and cooling rate were kept at 0.5°C/min. The procedure for impregnating the aluminosilicate powders was the same as the silica case. 6.3% mole $Ba(OH)_2 \cdot H_2O$ was dissolved in distilled water and the solution was poured into 2g of the aluminosilicate powder in a small tube and stirred for about half an hour with a spatula. The mixture was oven dried and then heat-treated at 400°C for 3 hrs @ 0.5 °C/min. GC results obtained for the impregnated aluminosilicate powder thus prepared is shown in Fig. 6. Results show two peaks at 50°C, 70°C and 90°C indicating that the CO_2 gas was retained in the GC column filled with the aluminosilicate powders. The N_2 gas being inert is not adsorbed on the powders and elutes first while the CO_2 takes longer time due to adsorption on the aluminosilicate powder. The second peak corresponds to the ejection of CO_2 gas from the GC column. Moreover, we can observe that the adsorption and hence the separation is more pronounced as the temperature is lowered from 90°C to 50°C. This is evident from the increasing differential elution times for the two species. Thus at 50°C, CO_2 is retained for a longer time than at 90°C. Quantitative evaluation of the effect of temperature over the separation or adsorption of CO_2 is possible by means of the parameter ΔH (or Q) called the heat of adsorption. A higher Q-value indicates a stronger affinity of the powders towards the gas. The heat of adsorption can be determined from the differences in retention times between CO_2 and N_2.

$$k = \frac{t_R - t_0}{t_0} = \frac{\Delta t}{t_0} \tag{1}$$

where k is the capacity factor. Also,

$$k = \frac{RT}{V} n_{total} A \exp\left(\frac{-\Delta H}{RT}\right) = TA' \exp\left(\frac{-\Delta H}{RT}\right) \tag{2}$$

where V is the interstitial volume and n_{total} is the total number of adsorption sites.

$$A' = \frac{R}{V} n_{total} A \tag{3}$$

where A and A' are constants.

Colloidal Ceramic Processing

Using the above equations we can find the heat of adsorption from the retention times in Fig. 6. This value is calculated as 27 kJ/mole. Heat of

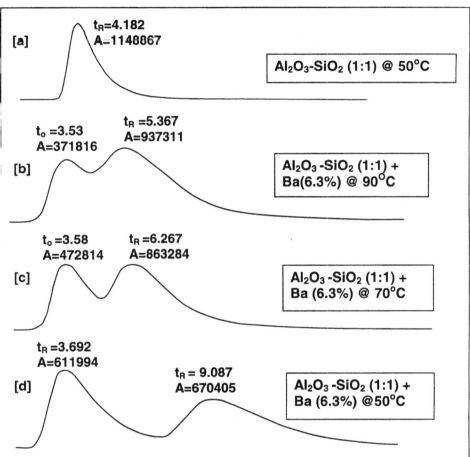

Fig.6 GC for SiO_2-Al_2O_3 with and without BaO (6.3%) obtained at 90°C, 70°C, and 50°C. [a] shows no adsorption of CO_2 even with alumina in the absence of BaO. [b], [c] and [d] are GC results for SiO_2-Al_2O_3 with 6.3% BaO. The two peaks represent the adsorption of CO_2 and give the elution times, t_0 and t_R, of N_2 and CO_2 respectively. The overlap of the two peaks is less pronounced at 50°C than at 70°C and 90°C. Flow rate is 1 ml per 36 seconds.

dsorption for pure alumina is cited in literature to be 78 kJ/mole.[6] The lowering f the heat of adsorption is thought to be due to the presence of silica. The results ı Fig. 6 showed that alumina is needed to adsorb CO_2 gas. More systematic udies on the effects of concentrations of alumina and BaO are needed and will e reported in future work.

CONCLUSION

It is shown that silica and Ba-doped silica do not adsorb CO_2 and, and thus cannot separate the CO_2 and N_2. Similar results were obtained with silica-alumina powders without any impregnated BaO. This may be because the alumina added to silica is not sufficient to adsorb the CO_2 gas. However, when the aluminosilicate powders are impregnated with BaO, we observe significant adsorption of CO_2. The addition of both alumina and BaO are crucial to bring about the affinity to the silica, and thus cause separation of CO_2 from N_2.

REFERENCES

[1] R. W. Baker, "Membrane Transport Theory"; pp. 15-86 in *Membrane Technology and Applications*, Edited by R. Esposito and S. Melvin, McGraw Hill. New York, 2000.

[2] P. Pandey and R. S. Chauhan, "Membranes for Gas Separation", *Progress in Polymer Science,* **26**, 853-893, 2001.

[3] W. J. Koros and R. Mahajan, "Pushing the Limits on Possibilities for Large Scale Gas Separation: Which Strategies?" *J. Membrane Science,* **175**, 181-196 2000.

[4] W. J. Koros and G. K. Fleming, "Membrane-based Gas Separation", *J Membrane Science,* **83**, 1-80, 1993.

[5] R. S. A. de Lange, J. H. A. Hekking, K. Keizer, and A. J. Burggraaf "Polymeric-Silica-Based Sols for Membrane Modification Applications: Sol-Gel Synthesis and characterization with SAXS", *J. Non-Crystalline Solids*, **191**, 1-16 1995.

[6] T. Horiuchi, H. Hidaka, T. Fukui, Y. Kubo, M. Horio, K. Suzuki, and T Mori, "Effect of Added Basic Metal Oxides on CO_2 Adsorption on Alumina a Elevated Temperatures", *Applied Catalysis A: General*, **167,** 195-202, 1995.

STUDY OF MECHANISM OF PYROCHLORE-FREE PMN-PT POWDER USING A COATING METHOD

Huiming Gu, Wan Y. Shih, and Wei-Heng Shih
Department of Materials Science & Engineering,
Drexel University,
Philadelphia, Pennsylvania 19104

ABSTRACT

The reasons why coating method can prevent the formation of pyrochlore phase in one-step calcination of perovskite phase PMN powders were studied. Using x-ray diffraction, DTA, and emission analysis, it was found that several reactions occurred during the calcination before 500°C. These reactions produce small pyrochlore phase particles. The small pyrochlore particles and the well mixing of chemical species provided by the coating method decreases the diffusion distance necessary for the transformation reaction from pyrochlore to perovskite phase.

INTRODUCTION

Lead magnesium niobate (PMN) and lead magnesium niobate-lead titanate (PMN-PT) are widely used relaxor ferroelectric materials because of their superior room temperature dielectric and electrostrictive properties.[1,2] However, the fabrication of PMN-PT is difficult compared with the traditional ferroelectric materials such as $BaTiO_3$ or lead zirconate titanate (PZT). For, PMN-PT, it is difficult to eliminate the low dielectric-constant pyrochlore phase during calcination. The most widely used method to avoid this problem is the two-step calcination columbite method introduced by Swartz and Shrout in 1982.[3] Since then, many one-step calcination methods were found such as sol-gel,[4,5,6] solution processes,[7,8] co-precipitation,[9] soft mechanochemical pulverization,[10] $Mg(NO_3)_2$ mixing,[11] and thermal spray methods.[12] These methods were based on the principle of optimizing the powder characteristics including particle size, specific surface area, reactivity of raw materials, or high-energy milling. Recently we discovered a one-step calcination method by colloidally coating $Mg(OH)_2$ on Nb_2O_5.[13] Compared with other one-step calcination method, our method is unique in that it does not require special raw materials and equipment.

Besides researching on the PMN-PT processing, efforts had been devoted to understand the reaction mechanism and kinetics in PMN-PT processing in the past decades. Inada[14] first proposed the following reaction sequence for the calcination of PbO, MgO, and Nb_2O_5 mixture:

$$3PbO + 2Nb_2O_5 \xrightarrow{500-600^\circ C} Pb_3Nb_4O_{13}(\text{Cubic Pyrochlore}) \qquad (1)$$

$$Pb_3Nb_4O_{13} + PbO \xrightarrow{600-700^\circ C} 2Pb_2Nb_2O_7(\text{Rhombohedral Pyrochlore}) \qquad (2)$$

$$Pb_2Nb_2O_7 + 1/3MgO \xrightarrow{700-800^\circ C} Pb(Mg_{1/3}Nb_{2/3})O_3(\text{Perovskite}) + 1/3Pb_3Nb_4O_{13} \ (3)$$

In this study, two intermediate phases, cubic pyrochlore $Pb_3Nb_4O_{13}$ and rhombohedral pyrochlore $Pb_2Nb_2O_7$, were produced. Later, Bouquin etc.[15] proposed a different reaction sequence:

$$3PbO + 2Nb_2O_5 \xrightarrow{500-600^\circ C} Pb_3Nb_4O_{13} \ (\text{Cubic Pyrochlore}) \qquad (4)$$

$$Pb_3Nb_4O_{13} + 2PbO \xrightarrow{630-690^\circ C} Pb_5Nb_4O_{15} \ (\text{Rhombohedral Pyrochlore}) \qquad (5)$$

$$Pb_5Nb_4O_{15} + PbO \xrightarrow{700-730^\circ C} 2Pb_3Nb_2O_8 \ (\text{Tetragonal Pyrochlore}) \qquad (6)$$

$$Pb_3Nb_2O_8 + MgO \xrightarrow{750-800^\circ C} 3Pb(Mg_{1/3}Nb_{2/3})O_3 \ (\text{Perovskite}) \qquad (7)$$

Most other studies showed only one pyrochlore phase was formed during the reaction with the following possible formulas: $Pb_{1.83}Nb_{1.71}Mg_{0.29}O_{6.39}$, $Pb_2Mg_{0.32}Nb_{1.87}O_7$, $Pb_{1.86}Nb_{1.76}Mg_{0.24}O_6$, $Pb_{2.25}Nb_{1.79}Mg_{0.27}O_7$, $Pb_2Nb_{1.75}Mg_{0.25}O_{6.62}$, $Pb_2Nb_{1.33}Mg_xO_{5.33+x}(0<x<0.66)$, $Pb_3(Mg_{1-x}Nb_{2+x})O_{9+3x/2}(0<x<0.625)$, $Pb_{(3+3x/2)/2}(Mg_xNb_{2-x})O_{0.65}(0<x<0.5)$, $Pb_{1.83}Mg_{0.29+x}Nb_{1.71-x}O_{6.39-1.5x}(0.1<x<0.522)$, and $Pb_{2-x}(Mg_{0.286}Nb_{1.714})O_{6.571-x}(0<x<0.286)$.[16]

The common features of these studies are that the pyrochlore phase is composed mainly of PbO and Nb_2O_5. Small amount of MgO may appear in the pyrochlore lattice. The ratio of Pb:Nb in these pyrochlore phases are less than the ratio of Pb:Nb=1.5 in the perovskite phase. In essence, the transformation reaction from the pyrochlore phase to the perovskite phase is mainly a process of MgO diffusing into the lattice of pyrochlore.

There were also several kinetic studies about the influence of processing on the percentage of perovskite phase formed. Lejeune and Boilot[17] found that improving the reactivity and dispersibility of MgO by using $MgCO_3$ as a reactant and acetone as ball-milling media could increase the percentage of perovskite phase formed. Calcination in O_2 atmosphere also improves the percentage of perovskite phase. Multiple calcination cycles can improve the perovskite percentage but high temperature calcination for long time will decrease the perovskite phase because of PbO loss by evaporation. Desgardin etc.[18] found that the uniform distribution of MgO was critical to the perovskite phase formation.

In this study, we investigated the reaction mechanism of perovskite phase formation. Our study focused on why our coating method can prevent the formation of pyrochlore phase by using micron-size Nb_2O_5 and PbO.

EXPERIMENTAL PROCEDURE

Starting materials used in this study were Nb_2O_5 (99.9% Aldrich), $PbTiO_3$ (99.+%, Aldrich), PbO (99.9+%, Aldrich), $Mg(NO_3)_2 \cdot 6H_2O$ (99%, Aldrich), MgO (99%, Aldrich) and ammonium hydroxide (5.08N Aldrich). To obtain the coated powder, 0.105 mole of $Mg(NO_3)_2 \cdot 6H_2O$ was dissolved in 500 ml distilled water followed by the addition of 0.1 mole of dispersed Nb_2O_5 powder suspension in the solution. This mixture was then stirred and ultrasonicated for ten minutes to break up the Nb_2O_5 agglomerates. At this point, the suspension pH was between 5 and 6. In order for $Mg(OH)_2$ to precipitate on the Nb_2O_5 surface, ammonium hydroxide (5.08 N) was added drop-wise into the mixture until the pH reached 10 and the mixture was kept stirring for 30 minutes. During this stage, $Mg(OH)_2$ was precipitated and coated on the surface of Nb_2O_5. The corresponding amount of PbO and 0.033 mole of $PbTiO_3$ (PT) then were added to the suspension. The suspension was ultrasonicated for another 10 minutes. After the mixture was stirred for an hour, it was dried by rotary evaporation. The dried powders were ball milled in isopropyl alcohol for 20 hours and rotary evaporated. The resulting powders were called coated powder. Perovskite phase PMN powders were produced from the coated powder after 2 hours' of calcinations at 900°C.

To study the reactions, differential thermal analysis (DTA, Model Universal V2.3C, TA Instruments, Inc., Newcastle, DE) and emission analysis were done on the coated powder. The Quantachrome NOVA 2200 was used to measure the specific surface area of powders. Scanning electron microscopy (SEM, Model 1830, Amray, Bedford, MA) was used to examine the fracture surface of pressed samples. X-ray diffractometry (XRD, Model D500, Siemens, Madison, WI) was used to identify the phase.

As will be discussed in the Results and Discussions part of this paper, nanosize MgO (92 nm) and Nb_2O_5 (35 nm) are also used. They were prepared by wet-chemical method. For nanosize MgO preparation, 0.105 mol $Mg(NO_3)_2 \cdot 6H_2O$ (99%, Aldrich) was first dissolved in 500 ml distilled water. While keeping the solution stirring, ammonium hydroxide (5.08N, Aldrich) was added dropwise until pH of the solution reached 10 resulting in the precipitation of $Mg(OH)_2$. The suspension was rotary evaporated and heat-treated at 500°C for 1 hour. The product was MgO powder of 92 nm in diameter. To prepare nanosize Nb_2O_5, 0.1 mole niobium oxalate (99%, Aldrich) was dissolved in 500 ml of distilled water. 30ml ammonium hydroxide (5.08N, Aldrich) was added dropwise while keeping the solution stirring. Precipitates were collected by centrifugation and followed by 80°C evaporation. Finally, it was heat-treated at 500°C for 3 hours. The product was Nb_2O_5 with 35 nm in diameter.

RESULTS AND DISCUSSIONS

Figure 1 shows the results of differential thermal analysis (DTA) and emission analysis of $Mg(OH)_2$-coated Nb_2O_5 powders plus PbO powders. The DTA result (Figure 1(a)) showed that there were reactions at <200°C, 330°C, 434°C, 483°C, 569°C, and 807°C during calcination. Emission analysis results (Figure 1(b)) showed that three different kinds of gas with molecular weight 18, 30, and 46 were evaporated during calcination. Since only elements of Nb, Mg, Pb, Ti, O, N, and H existed in the system, the only possible gas emissions were H_2O or NH_3 for molecular weight 18, NO for 30, and NO_2 for 46.

To analyze the reactions found in the DTA results, x-ray diffraction was performed on coated powders calcined at different temperatures. These samples were prepared by first heat-treating the as-synthesized coated powder to the desired temperature at the rate of 5°C/min. After holding at the temperature for 1 hour, the powders were cooled to room temperature at the rate of 10°C/min. Figure 2 shows the results of the x-ray diffraction. It can be seen that the as-synthesized powder was a mixture of Nb_2O_5 and $3Pb(NO_3)_2 \cdot 7PbO$. The PbO initially added has become $3Pb(NO_3)_2 \cdot 7PbO$ after coating, evaporation, and ball-milling. Since the reactant used to precipitate $Mg(OH)_2$ was $Mg(NO_3)_2$, it is possible for the $(NO_3)^-$ ions from the $Mg(NO_3)_2$ to react with PbO and forming $3Pb(NO_3)_2 \cdot 7PbO$ in the solution or during evaporation. Reaction peaks below 200°C were observed in Figure 1(a). Considering there was a large amount of H_2O or NH_3 evaporating revealed in Figure 1(b) in this temperature range, we attributed these reaction peaks to the evaporation of bound water and NH_3. In Figure 2(a), no peaks of Mg-compound were found in the diffraction pattern below 300°C and there was only a barely noticeable MgO peak in the diffraction pattern at 350°C. This is consistent with the fact that MgO only occupy 4.2 wt% in $Pb(Mg_{1/3}Nb_{2/3})O_3$ and Mg-compound is likely in the amorphous state.

In order to investigate the phase change of Mg-compound during calcination, a solution of $Mg(NO_3)_2$ with the same concentration as that used in the coating method were prepared. Similar to the coating method, ammonium hydroxide was added to precipitate $Mg(OH)_2$. Unlike the coating method, no Nb_2O_5 powder was added to coat $Mg(OH)_2$ on and no PbO were added after precipitation. In this way, the pure Mg-compound coating substance was prepared. The Mg-compound was heat-treated to several temperatures and the x-ray diffractions were performed. Figure 3 shows the x-ray diffraction pattern of the Mg-compound calcined to 300°C, 350°C, and 400°C. The pattern showed that, at 300°C, Mg-compound was composed mainly of $Mg(OH)_2$ with a small amount of $Mg_3(OH)_4(NO_3)_2$. At 350°C, the diffraction pattern showed $Mg(OH)_2$ disappeared and MgO was the dominant phase. This means that coating precipitation mainly consisted of $Mg(OH)_2$. The $Mg(OH)_2$ coating will decompose to MgO between 300°C and 350°C. This temperature range matched well with the $Mg(OH)_2$ decomposition temperature of 330°C from physical chemistry handbook. So we conclude that the reaction at 330°C detected by DTA (Figure 1 (a)) is the transformation reaction from $Mg(OH)_2$ to MgO with the following equation:

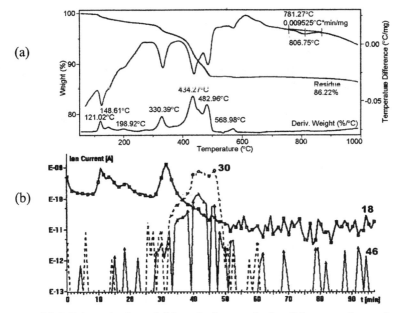

Figure 1. (a) DTA analysis and (b) emission analysis of the coated powder.

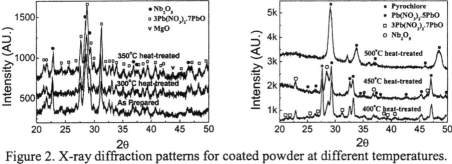

Figure 2. X-ray diffraction patterns for coated powder at different temperatures.

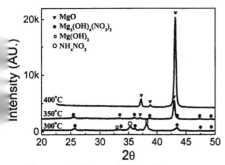

Figure 3. X-ray diffraction pattern
of Mg-compound.

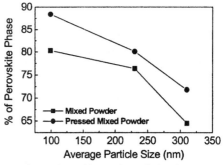

Figure 4. Effect of particle size and
particle configuration.

$$Mg(OH)_2 \xrightarrow{330°C} MgO + H_2O \tag{8}$$

The emission peak of gas with molecular weight 18 at this temperature range in Figure 2 (b) confirms this reaction. The small amount of $Mg_3(OH)_4(NO_3)_2$ did not decompose until 400°C. To investigate the particle size of MgO from the $Mg(OH)_2$ decomposition, specific surface area of MgO from $Mg(OH)_2$ precipitation calcined at 400°C and 500°C respectively were measured. The results were listed in Table 1. The particle size in the table was calculated from the specific surface area using equation $D = 6/\rho(SSA)$ where ρ is the theoretical density, SSA is the specific surface area, and D is the calculated particle size. The results of calculated particle size suggest that the MgO made from $Mg(OH)_2$ decomposition was about 100 nm in size. This size is much smaller than the commercial MgO, which is in the range of micrometer.

Table 1. Specific surface area of MgO from $Mg(OH)_2$ decomposition

	Specific Surface Area	Particle size
400°C heat treatment	9.3 m²/g	180 nm
500°C heat treatment	18.3 m²/g	92 nm

Figure 2 (b) was the x-ray diffraction pattern of coated powders calcined at temperatures 400°C, 450°C and 500°C. It showed that the powder was $3Pb(NO_3)_2 \cdot 7PbO$ plus Nb_2O_5 at 400°C, $Pb(NO_3)_2 \cdot 5PbO$ plus Nb_2O_5 at 450°C, and pyrochlore at 500°C respectively. In the same temperature range, peaks of NO and NO_2 were observed from the emission analysis (Figure 1(b)). As a result, the following equations are attributed to the 434°C and 482°C reactions found in DTA (Figure 1(a)).

$$3(3Pb(NO_3)_2 \cdot 7PbO) \xrightarrow{434°C} 5(Pb(NO_3)_2 \cdot 5PbO) + 8xNO + 8(1-x)NO_2 + 2(2x+1)O_2 \tag{9}$$

$$Pb(NO_3)_2 \cdot 5PbO + 6aNb_2O_5 + 6bMgO \xrightarrow{482°C}$$
$$6PbNb_{2a}Mg_bO_{1+5a+b} (\text{pyrochlore}) + 2xNO + 2(1-x)NO_2 + (x+1/2)O_2 \tag{10}$$

where $0 \leq x \leq 1$, $1/3 \leq a \leq 1$, $0 \leq b < 1/3$. The range of a and b were given to indicate all possible pyrochlore formulas discussed in the Introduction.

The specific surface area of coated-powder calcined at 500°C was measured to be 6.39 m²/g, corresponding to a diameter of 132 nm. The corresponding surface area and particle size for powders obtained from regular oxide powder mixing calcined at the same temperature was 1.30 m²/g and 649 nm.

The reactions above 500°C are related to the transformation from pyrochlore to perovskite phase. Several papers have studied these reactions[14,15,16] and we will not discuss them here. The general conclusions of these studies were that pyrochlore was an MgO-deficient phase. The transformation is a process of MgO diffusing into the pyrochlore lattice.

So far we have shown that the pyrochlore phase formed after 500°C calcination for both coated power and regular oxide mixed power. The differences between the pyrochlore phase from these two methods are: (1). The particle size of pyrochlore phase particle from the coating method was much smaller than that from the oxide mixing method. (2). The MgO in the coating method came from the decomposition of $Mg(OH)_2$. The particle size of MgO formed from decomposition is smaller. (3). Initially the $Mg(OH)_2$ was coated on Nb_2O_5 so the configuration of MgO after 500°C calcination was different from that of oxide mixing.

In order to show the influence of particle size and particle configuration to the pyrochlore-perovskite phase transformation, three types of powder samples were prepared. The first type of sample was a regular MgO, Nb_2O_5, and PbO powder mixture ball-milled for 2 hrs. The average particle size was 320 nm. The second type of sample was a regular MgO, Nb_2O_5, and PbO powder mixture ball-milled for 48 hrs. The average particle size was 240 nm. The third type of sample was a nanosize MgO (92 nm), nanosize Nb_2O_5 (35 nm), and regular PbO powder mixture ball-milled for 48 hrs. The average particle size was 99 nm. Two groups of samples were made from each of the three types of power. One group was made by heat-treating the powders at 900°C for two hrs and another group was pieces made by heat-treating 100 MPa compacted powders at 900°C for 2 hrs. The difference between these three types of powders was the particle size. The difference between the two groups of samples for each type was particles configuration. Figure 4 shows the percentage of pyrochlore phase formed in each sample. This result showed that smaller particle size and close compact help the transformation from pyrochlore phase to perovskite phase. This meant that particle size and configuration have a significant effect on the transformation from pyrochlore phase to perovskite phase. This figure also showed that the small particle size itself was not sufficient to prevent the formation of pyrochlore phase.

As illustrated in the Introduction, composition distribution, especially the MgO distribution, is a key factor that influences the transformation from the pyrochlore phase to the perovskite phase. The more uniform the distribution of MgO, the more complete the transform from the pyrochlore to the perovskite phase. Both particle size and configuration have an effect in influencing the composition distribution, which is a critical factor in completing the reaction. In order to compare these effects quantitatively, a simple simulation of randomly placing mixed oxide and coated powders that are either loose or pressed was performed.

The simulation was done on a two-dimensional 200×200 hexagonal close packed lattice. Each lattice point could be occupied by one particle as illustrated in Figure 5. For the case of mixed oxide powders, three different particles representing PbO, Nb_2O_5, MgO respectively with the mole ratio 3:1:1 were placed on the lattice randomly. For the case of coated powders, particles with coating were used to represent MgO and Nb_2O_5. The coating thickness was distributed randomly from 50% to 150% of average thickness to represent the non-uniformity of thickness in the coating. The coated particles and the PbO particles were randomly placed on the lattices. To represent the loose powder condition, only 25% of the lattices were occupied with particles. On the other hand, all lattices were occupied to represent the compacted powders. Figure 5 shows the snapshots of the simulation. In the snapshot, one lattice point was randomly picked. Base on this lattice point, a hexagon was drawn around it. The compositions of components in this hexagon were calculated and compared with the overall composition of PbO, MgO, and Nb_2O_5. We define the shortest edge of this hexagon in which all the composition were within a 5 % variation of the overall composition as the compositional uniform radius. In essence, solid-state reaction was a process of inter-diffusion between components with each other to form a compositional uniform phase. Under certain temperature and reaction time, the shorter the diffusion distance, the more complete the reaction. This compositional uniform radius we defined was the minimum range for compositional uniformity. Each component would have to diffuse at least in the order of this distance to complete the reaction. So this compositional uniform radius is a parameter related to the diffusion distance. The larger the compositional uniform radius, the longer the components need to diffuse for reaction.

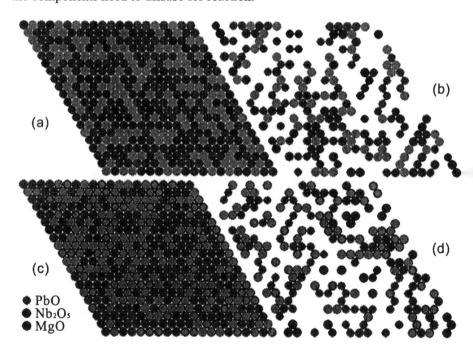

(a)

(b)

(c)

(d)

● PbO
● Nb₂O₅
● MgO

Figure 5. Simulation snapshots. (a) Pressed oxide powders, (b) loose oxide powders, (c) pressed coated powders (d) loose coated powders.

Figure 6 shows the simulation results for the compositional uniformity radius with error bars. The number was obtained from the average of 200 independent runs. The unit of radius was the average particles size initially added. The larger the initial average particle size, the longer the diffusion distance, and the more difficult for the reaction to complete. This could explain the higher perovskite yield percentage in Figure 4 for smaller particle size. Figure 6 also showed that the compacted piece has a shorter diffusion distance than the loose powder. This could explain the higher perovskite yield percentage in Figure 4 for pressed pieces. Figure 6 also showed that the coated powders have a shorter diffusion distance than the mixed oxide powders. This is the main factor that prevents the appearance of pyrochlore phase in the coating method.

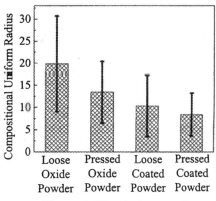

Figure 6. Simulation results.

CONCLUSION

In this paper we showed that several reactions occurred before 500°C during calcinations for the Mg(OH)$_2$-coated Nb$_2$O$_5$ powders. These reactions produce very small pyrochlore particles and maintain a configuration of MgO particle mixed uniformly with the Nb$_2$O$_5$ particles. The smaller pyrochlore particle size and the better mixing configuration decrease the diffuse distance in reaction and thus help complete the transformation from pyrochlore to perovskite phase near 900°C for the coated powders.

REFERENCES

[1] V. A. Bonner, E. F. Dearbon, J. E. Geusic, H. M. Marcos, and L. G. Van Uitert, "Dielectric and Electro-optic Properties of Lead Magnesium Niobate," *Appl. Phys. Lett.*, **10** [5] 163-165 (1967).

[2] L. E. Cross, "Relaxor Ferroelectrics," *Ferroelectrics*, **151** 305–20 (1994).

[3] S. L. Swartz and T. R. Shrout, "Fabrication of Perovskite Lead Magnesium Niobate," *Mater. Res. Bull.*, **17** 1245–50 (1982).

[4] P. Ravindranathan, S. Komareni, A. S. Bhalla, and R. Roy, "Synthesis and Dielectric Properties of Sol–Gel-Derived 0.9Pb(Mg$_{1/3}$Nb$_{2/3}$)O$_3$–0.1PbTiO$_3$ Ceramics," *J. Am. Ceram. Soc.*, **74** [12] 2996–99 (1991).

[5] F. Chaput, J. P. Boilot, M. Lejeune, R. Papiernik, and L. G. Hubert Pfalzgraf, "Low-Temperature Route to Lead Magnesium Niobate," *J. Am. Ceram. Soc.*, **72** [8] 1355–57 (1989).

[6] Y. Narendar and G. L. Messing, "Kinetic Analysis of Combustion Synthesis of Lead Magnesium Niobate from Metal Carboxylate Gels," *J. Am. Ceram. Soc.*, **80** [4] 915–24 (1997).

[7] J. P. Lee, J. Lee, S. Kang, and H. Kim, "Synthesis of $Pb(Mg_{1/3}Nb_{2/3})O_3$ Powder by Solvent Evaporation and Its Dielectric Properties," *J. Korean Ceram. Soc.*, **33** [1] 17–24 (1996).

[8] M. Lejeune and J. P. Boilot, "Formation Mechanism and Ceramic Process of the Ferroelectric Perovskites: $Pb(Mg_{1/3}Nb_{2/3})O_3$ and $PbFe_{1/2}Nb_{1/2}O_3$," *Ceram. Int.*, **8** [3] 99–104 (1982).

[9] W. B. Ng, J. Wang, S. C. Ng, and L. M. Gan, "Processing and Characterization of Microemulsion-Derived Lead Magnesium Niobate," *J. Am. Ceram. Soc.*, **82** [3] 529-36 (1999).

[10] J. G. Baek, T. Isobe, and M. Senna, "Synthesis of Pyrochlore-Free 0.9Pb $(Mg_{1/3}Nb_{2/3})O_3$–$0.1PbTiO_3$ Ceramics via a Soft Mechanochemical Route," *J. Am. Ceram. Soc.*, **80** [4] 973–81 (1997).

[11] K. R. Han, S. Kim, and H. J. Koo, "New Preparation Method of Low-Temperature-Sinterable Perovskite-Free $0.9Pb(Mg_{1/3}Nb_{2/3})O_3$–$0.1PbTiO_3$ Powder and Its Dielectric Properties," *J. Am. Ceram. Soc.*, **81** [11] 2998–3000 (1998).

[12] A. L. Costa, C. Galassi, and E. Roncari, "Direct Synthesis of PMN Samples by Spray-drying," *J. Euro. Ceram. Soc.*, **22** 2093–2100 (2002).

[13] H. M. Gu, W. Y. Shih, and W.-H Shih, "Single-Calcination Synthesis of Pyrochlore-Free $0.9Pb(Mg_{1/3}Nb_{2/3})O_3$–$0.1PbTiO_3$ and $Pb(Mg_{1/3}Nb_{2/3})O_3$ Ceramics Using a Coating Method," *J. Am. Ceram. Soc.*, **86** [2] 217–21 (2003).

[14] M. Inada, "Analysis of the Formation Process of the Piezoelectric PCM Ceramics," *Jpn. Natl. Tech. Rept.*, **27** [1] 95-102(1977)

[15] O. Bouquin, and M. Lejeune, "Formation of the Perovskite Phase in the $PbMg_{1/3}Nb_{2/3}O_3$-$PbTiO_3$ System," *J. Am. Ceram. Soc.*, **74** [5] 1152-56 (1991)

[16] J.P. Guha, "Reaction chemistry and subsolidus phase equilibria in lead-based relaxor systems: Part I - Formation and stability of the perovskite and pyrochlore compounds in the system PbO-MgO-Nb_2O_5", *J. Mater. Sci.*, **34** 4985-4994 (1999)

[17] M. Lejeune, and J. P. Boilot, "Influence of Ceramic Processing on Dielectric Properties of Perovskite Type Compound: $Pb(Mg_{1/3}Nb_{2/3})O_3$," *Ceram. Int.*, **9** [4] 119-22 (1993)

[18] G. Desgardin, A. Bali, and B. Raveau, "Ceramiques Composites A Basae De Perovskite et Pyrochlores Au Plomb PZN $(PbZn_{1/3}Nb_{2/3})O_3$: PFN $(PbFe_{1/2}Nb_{1/2}O$ et PMN $(PbMg_{1/3}Nb_{2/3})O_3$, {pir Condensateurs Multicouches a Haute Constante Dielectrique," *Mater. Chem. Phys.*, **8**, 469-91 (1993)

MICROMECHANICAL TESTING OF TWO-DIMENSIONAL AGGREGATED COLLOIDS

Sarunya Promkotra and Kelly T. Miller
Department of Metallurgical and Materials Engineering
Colorado School of Mines
Golden, CO 80401

ABSTRACT
A novel micromechanical testing technique has been developed which allows the mechanical properties of small clusters of aggregated particles to be determined. Model two-dimensional aggregated colloids were produced, composed of 4 μm particles floating on a water-glycerol substrate. These aggregated colloids display both elastic and plastic behavior. Both the elastic modulus and compressive yield stress increased by over two orders of magnitude, as the area fraction of particles was increased. The observed moduli and yield stresses are also highly variable and depend upon the specific microstructure of the cluster.

INTRODUCTION
The compressive yield stress of flocculated suspensions is a material property of the suspension, which can be used to predict the suspension behavior during such processing operations as pressure filtration or slip casting [1-4]. The compressive yield stress increases with increasing interparticle attractive force, decreasing particle size, and increasing volume fraction, and depends upon suspension microstructure [5]. The scaling behavior with respect to particle volume fraction in particular remains poorly understood. The yield stress increases by many orders of magnitude as volume fraction increases, and has been described by exponential laws, power laws, and other relations [5, 6]. The shear yield stress scales similarly to the compressive yield stress with volume fraction, but is usually one to two orders of magnitude lower in value [3]. Similar behavior is also seen during the dry pressing of ceramic powder. The ability to predict these yield stresses for a particular suspension remains poor.

Flocculated suspensions have been extensively studied by light scattering experiments [7] and computer simulations [8]. Few experiments, however, have directly examined the evolving microstructure during the yield of colloidal aggregates. In this paper, the deformation properties of floating two-dimensional colloidal particle aggregates are directly studied using a newly developed experimental technique, which combines digital video microscopy with a novel micromechanical testing method [9, 10]. Digital video microscopy is used to directly observe and quantify the particle movement during compression, and micromechanical testing was used to measure the deformation forces required for particle rearrangement at the microstructural level.

EXPERIMENTAL PROCEDURE

Formation of Monoparticulate Layer at Air-Liquid Interface

Aggregated floating two-dimensional colloids of polystyrene particles were prepared on substrates composed of water-glycerol solutions. Monodispersed polystyrene spheres with sulfate functional groups (Interfacial Dynamics Corporation, Portland OR, USA), 4 μm in diameter, were suspended in a light alcohol (such as methanol, ethanol) at 0.1% solid fraction. In this suspension, particles were weakly flocculated, so they were placed in an ultrasonic bath for 30-60 minutes in order to break up doublets in suspension. The substrate solutions were composed of 50-70 wt% of glycerol dissolved in water. The substrate solution was placed in 15 x 15 mm chamber slide (Lab-Tek, Nalge Nunc International). The substrate solution was 2.67 mm deep, giving a total volume of 600 μL [10].

10-20 μL of the polystyrene suspension was deposited on the water-glycerol solution substrate (Figure 1). The particles would rapidly aggregate at the air-liquid interface after the evaporation of the alcohol, which took 20-60 s. Large floating networks of aggregated particles would develop in about approximately one hour by connecting each of the isolated small aggregates.

To induce aggregation, the acidity of the substrate solution was controlled at pH ~2.5 by addition of hydrochloric acid. Without the addition of the acid, the

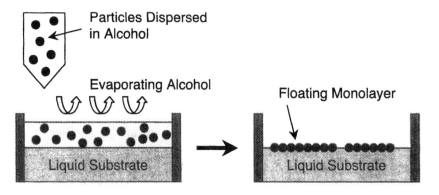

Figure 1. Side-view illustration of the formation of a 2D aggregated colloid.

particles trapped at the substrate-air interface would remain dispersed. We have also observed that aggregation can be induced through salt additions. For example, adding 0.10 M of an electrolyte solution such as sodium chloride or magnesium chloride screened the repulsive electrostatic interactions and induced aggregation of polystyrene microspheres. However, only pH induced aggregates were mechanically tested in this study.

This experimental technique can produce a variety of aggregated network morphologies. The completed aggregation always develops from small clusters connecting to others to growing a large network. The fractal dimension of aggregate could be varied between 1.55-1.85, depending upon the substrate conditions. Generally, 2D fractal cluster networks are formed at pH 1.5-2.5 for alcohol-polystyrene suspensions. A typical aggregated microstructure is shown in Figure 2.

Images of aggregation and loading experiments were obtained using an inverted transmission optical microscope (Carl Zeiss, Axiovert S100) and a grayscale 1/2-inch CCD-video camera (Cohu, Inc. Model 4910 series). The images were immediately digitized with a frame grabber card (Scion LG3) in a personal computer. Video was analyzed with the public domain image analysis programs NIH Image and ImageJ (written by Wayne Rasband at the U.S. National Institutes of Health).

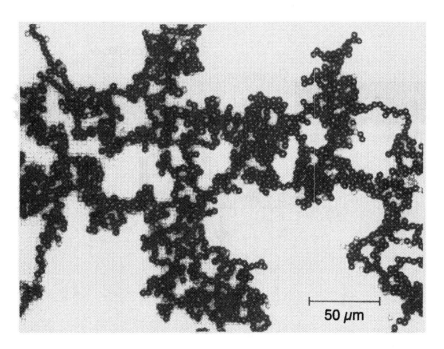

Figure 2. Morphology of aggregated 4 μm polystyrene network.

Micromechanical Testing

The micromechanical testing apparatus is schematically shown in Figure 3. This technique is based upon using micromanipulators which hold glass fiber cantilevers to load the aggregate, while simultaneously tracking the displacement of the cantilevers using digital video microscopy [9].

The cantilevers used here were glass fibers of 30 μm in diameter, bent into an L-shape using a microforge (Narishige MF-900). These were then glued onto a 1 mm glass rod and held with the micromanipulators (Figure 3). On the left side of the apparatus, a glass fiber was attached to a relatively low-precision joystick micromanipulator (Narishige MN-151), which remained stationary during the experiments. Displacement of this micromanipulator was used to measure the force applied to the sample. A second L-shaped glass fiber on the right side of apparatus was attached to a high precision micromanipulator (Eppendorf, Patchman 5171). This micromanipulator provides movement of 320 nm/sec in 160 nm steps on along the x, y, and z axes.

To align the apparatus, the micromanipulators were first adjusted to the height of the liquid substrate. The high precision micromanipulator was then used to capture a floating aggregate, which was then maneuvered so that both glass fibers entrapped the 2D cluster. After capture, the high precision manipulator was used to load the sample.

The force applied to the cluster was calculated from the measured movement of the stationary glass fiber cantilever, through the relation $F = k\ \delta$, where k is the spring constant of the glass fiber cantilever and δ is the displacement from the

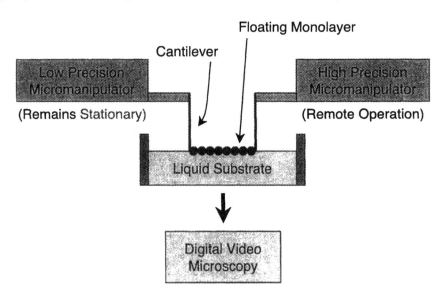

Figure 3. Schematic side-view illustration of the *in-situ* compressive loading apparatus.

Colloidal Ceramic Processing

fiber equilibrium position. The spring constant k of the glass fiber was calculated from the measured fiber geometry, using the relation for a cylindrical cantilever:

$$k = \frac{3\pi E r^4}{4L^3} \tag{1}$$

where E is the elastic modulus (70 GPa), r is the fiber radius (typically 12.5 - 15.0 μm in these experiments), and L is the fiber length (which was varied from 9 - 15 mm). For the typical fiber dimensions used in this experiment, $L = 9.5$ mm and $r = 15.3$ μm, the spring constant k is 1.0×10^{-2} N/m.

The strain on the sample was determined from the relative movement of the left and right fibers during the experiment, normalized by the initial cluster length (which was given by the initial distance between the fibers).

RESULTS AND DISCUSSION

Figure 4 shows the loading cycle for a typical experiment. The force applied to this cluster was calculated from the displacement of the left glass fiber, which in this case had spring constant k of 9.33×10^{-3} N/m. In this test, the loading was begun at t = 25 s by moving the high precision micromanipulator at 320 nm/s. Movement was stopped at t = 73.5 seconds, and finally the movement of the

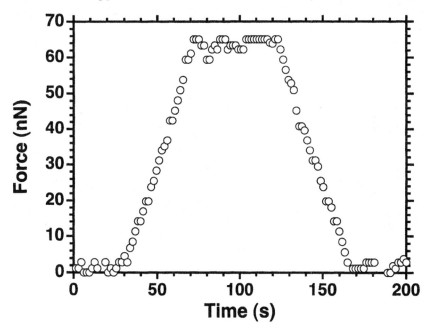

Figure 4. Force as a function of time during cluster compressive loading.

manipulator was reversed at t = 123 s until the sample was completely unloaded at t = 170 s. Note that the load remains constant while the manipulator was stopped; the sample was able to elastically support the applied load. The compressive loading and unloading were then repeated under the same conditions and similar results were obtained.

More typically, both elastic and plastic behavior are observed. Figure 5 shows a series of images from such a cluster deformation experiment. At low loads, the sample deforms elastically. At a critical load, however, the sample is observed to yield, and the cluster typically densifies and becomes stiffer. Figure 6 shows a load-strain curve of a sample which plastically densifies. The slope of the linear region increases upon unloading, indicating an increased elastic stiffness. The increased stiffness is retained upon subsequent reloading of the sample (Figure 7).

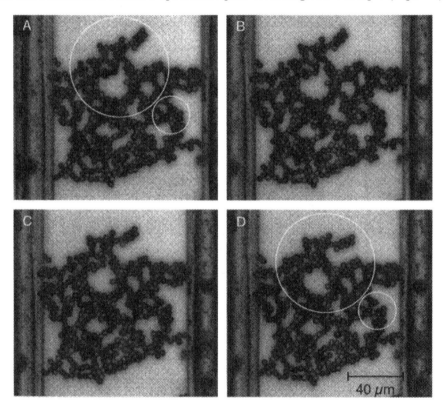

Figure 5. Elastic and plastic deformation of a cluster during compressive loading. (a) Unloaded sample. (b) Maximum elastic load. Although difficult to distinguish with the unaided eye, image analysis shows that this sample has a 1.1% elastic strain. (c) Sample during yield. (d) Sample after 4.6% plastic strain. Plastic deformation in the sample is concentrated in the white circled areas.

Colloidal Ceramic Processing

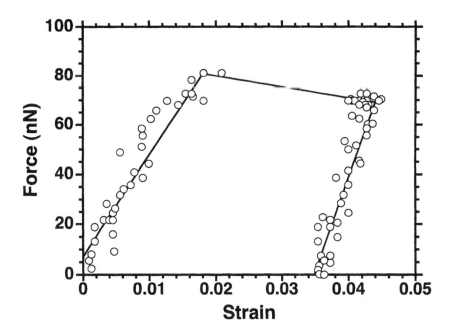

Figure 6. Load-strain curve of a plastically deforming cluster.

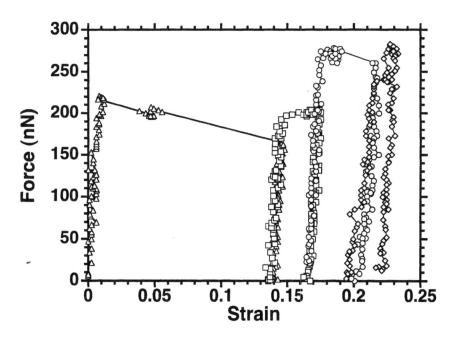

Figure 7. Load-strain cycling of a plastically deforming cluster.

Individual clusters were subjected to a series of cyclic loading-unloading experiments with increasing load levels. Figure 7 shows an example of the load-strain history for a single cluster. Note that yield stress and the slope of the linear elastic region tend to increase with the plastic strain and cluster density, although there is significant variability.

From the load-strain cycles of a series of clusters, elastic modulus (Figure 8) and yield stress (Figure 9) were calculated as a function of particle area fraction. In these curves, the behavior of individual clusters overlap, indicating that there is a characteristic cluster mechanical behavior. The behavior of the two dimensional colloids shows similar features to those seen in strongly aggregated three-dimensional macroscopic suspensions. The modulus and yield stress both increase by two orders of magnitude as the area fraction of particles is increased. In addition, the observed values are also highly variable. This is not, however, merely due to noise in the experimental data. It reflects the highly variable nature of the microstructure and the ability of specific arrangements of particles within the clusters to withstand the forces imposed upon them. Similar mechanisms are likely to be acting in macroscopic three dimensional systems.

CONCLUSIONS

A new micromechanical testing technique has been developed which allows direct measurements of the deformation properties of particle aggregates to be performed. Two dimensional aggregates of 4 μm polystyrene particles were used

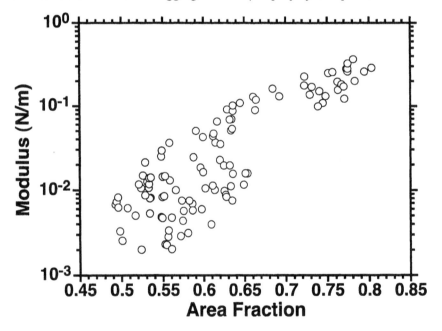

Figure 8. Cluster elastic modulus with particle area fraction.

Colloidal Ceramic Processing

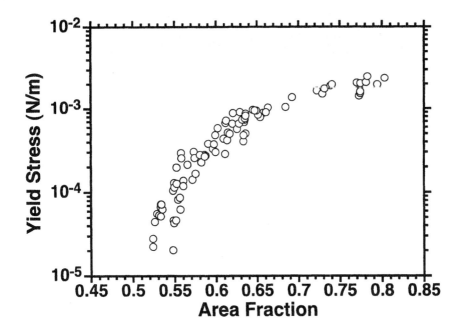

Figure 9. Cluster yield stress with particle area fraction.

as the model system. The aggregates displayed both elastic and plastic behavior. As in macroscopic three-dimensional systems, the measured elastic modulus and compressive yield stress increased by orders of magnitude as the concentration of particles was increased. The measured moduli and stresses were also highly variable, reflecting a dependence upon the specific small scale microstructural features of the sample.

ACKNOWLEDGEMENT

This work was supported by the National Science Foundation under CAREER Award # 9876135.

REFERENCES

[1]G. M. Channell and C. F. Zukoski, "Shear and Compressive Rheology of Aggregated Suspensions," *AIChE J.*, **43** [7] 1700-1708 (1997).

[2]L. Bergstrom, C. H. Schilling, and I. A. Aksay, "Consolidation Behavior of Flocculated Alumina Suspensions," *J. Am. Ceram. Soc.*, **75** [12] 3305-3314 (1992).

[3]R. Buscall, P. D. A. Mills, J. W. Goodwin, and D. W. Lawson, "Scaling Behaviour of the Rheology of Aggregate Particles Formed from Colloidal Particles," *J. Chem. Soc., Faraday Trans. 1*, **84** [12] 4249-4260 (1988).

[4]K. T. Miller, R. M. Melant, and C. F. Zukoski, "Comparison of the Compressive Yield Response of Aggregated Suspensions: Pressure Filtration, Centrifugation, and Osmotic Consolidation," *J. Am. Ceram. Soc.*, **79** [10] 2545-2556 (1996).

[5]G. M. Channell, K. T. Miller, and C. F. Zukoski, "Effects of Microstructure on the Compressive Yield Stress," *AIChE J.*, **46** [1] 72-78 (2000).

[6]M. Eberl, K. A. Landman, and P. J. Scales, "Scale-up Procedures and Test Methods in Filtration: A Test Case on Kaolin Plant Data," *Colloids Surf. A: Physicochem. Eng. Aspects*, **103** [1/2] 1-10 (1995).

[7]S. Tang, "Prediction of Fractal Properties of Polystyrene Aggregates," *Colloids Surf. A: Physicochem. Eng. Aspects*, **157**, 185-192 (1999).

[8]A. H. L. West, J. R. Melrose, and R. C. Ball, "Computer Simulations of the Breakup of Colloid Aggregates," *Phys. Rev. E*, **49** [5] 4237-4249 (1994).

[9]K. L. Eccleston and K. T. Miller, "Direct Measurement of Strongly Attractive Particle-Particle Interactions," pp. 33-38 in *Improved Ceramics Through New Measurements, Processing, and Standards (Ceramic Transactions v. 133)*, ed. by M. Matsui, S. Jahanmir, H. Mostaghaci, M. Naito, K. Uematsu, R. Wasche, and R. Morrell, American Ceramic Society (2002).

[10]S. Promkotra and K. T. Miller, "In Situ Observation of Rearrangement Mechanisms in Low Volume Fraction 2D Aggregated Colloids," pp. 39-47 in *Improved Ceramics Through New Measurements, Processing, and Standards (Ceramic Transactions v. 133)*, ed. by M. Matsui, S. Jahanmir, H. Mostaghaci, M. Naito, K. Uematsu, R. Wasche, and R. Morrell, American Ceramic Society (2002).

RHEOLOGY OF CERAMIC SLURRIES FOR ELECTRO-DEPOSITION IN RAPID PROTOTYPING APPLICATIONS

Navin Jose Manjooran, Seung-woo Lee, Georgios Pyrgiotakis, Wolfgang M Sigmund
Materials Science and Engineering
University of Florida
Gainesville, FL 32601

ABSTRACT

The rheology for alpha silicon carbide slurries in the presence of a charging agent with polybutadiene, polystyrene and LP1 as dispersants in an organic dispersing media, were investigated. The results were fit to the Krieger-Dougherty equation determining the maximum solids loading for the ceramic slurries. The maximum solids loading attainable calculated using the Krieger-Dougherty fit equation for slurries with polystyrene was 0.22 [η=12.8], with LP1 was 0.55 [η=5.41] and with polybutadiene was 0.69 [η=5.24]. Electro-deposition of these slurries yielded reversible deposition for slurries in the range of 5 vol% SiC. The optical thickness (deposited amount) of these layers sound promising for rapid prototyping applications, particularly the Electro-photographic Solid Freeform Fabrication (ESFF) technique.

INTRODUCTION

Most ceramic applications require slurries to have the highest possible solids loading and a low enough viscosity so that it can be easily poured. A high solid loading helps in reducing the sintering shrinkage. The maximum solid loading that is possible for a slurry can be calculated using the Krieger-Dougherty equation. The Krieger- Dougherty equation is generally used to give the relation between the relative viscosity and solids loading for hard sphere systems. The Krieger-Dougherty equation is given below (Equation 1) [1].

$$\eta_r = \left(1 - \frac{\phi}{\phi_m}\right)^{-[\eta]\phi_m} \tag{1}$$

where, ϕ is the volume fraction of solids, ϕ_m is the maximum solids loading or packing fraction, $[\eta]$ is the intrinsic viscosity and η_r is the relative viscosity. The objective of this article is to present the effect of volume fraction of solids on the deposition behavior for rapid prototyping applications, particularly for the Electro-photographic Solid Freeform Fabrication (ESFF) technique and the colloidal stability of these slurries using the Krieger-Dougherty fit for the different non-polar alpha silicon carbide slurries.

RAW MATERIALS AND SUSPENSION PREPARATION

6H-alpha- silicon carbide (Grade UF-15, H. C. Starck, Canada), cis-trans decahydronapthalene (Aldrich), polystyrene (Aldrich), polybutadiene (Aldrich), LP1 (Uniqema, Belgium) and charge controlling agent 7 (CCA7, Avecia-Inc) were the starting materials. The average molecular weights for polystyrene, polybutadiene and LP1 are ~230,000, ~ 5,000 and ~ 6,000, respectively. Silicon carbide (SiC) and CCA7 particle size were measured (Brookhaven instruments – zeta plus particle sizing) giving $d_{50}=0.52\pm0.02$ µm and $d_{50}=0.42\pm0.02$ µm, respectively. The specific surface area of SiC, measured by H.C. Starck using BET (AREAMETER II) N_2 adsorption is $A=15m^2/g$. Polystyrene, polybutadiene and LP1 are used as dispersants and their amounts are based on the weight percent of the dry SiC powder.

The slurries were prepared in the following way. The polymer dispersants were dissolved with measured amounts of decahydronapthalene while stirring on a hot plate. This is followed by the addition of measured amounts of CCA7 and the SiC powder. Then the mixture is placed in a misonix sonicator 3000 ultrasonic precursor for 4 hours. The generator provides high voltage energy pulses at 20 kHz. A titanium disruptor horn transmits and focuses oscillations of the piezoelectric crystals and produces the shearing and tearing action necessary for the slurry formation. This procedure is used for making all slurries.

Viscosity measurements were performed using a modular compact rheometer (MCR 300, Paar Physica) with a concentric cylinder system using the US200 universal software. The inner cylinder diameter is 27 mm. The shear flow measurements are operated at 298 K. The shear rate varies from 0.001 to 1000 s^{-1}. The temperature control unit is TEZ150P, which features peltier heating. The slurry is pre-sheared at 800 s^{-1} for one minute. The slurry is then kept stationary for 10 seconds to equilibrate. Then the measurement is carried out. The relative viscosity is the ratio of the viscosity of the suspension to the viscosity of decahydronapthalene at the same temperature.

Colloidal Ceramic Processing

DETERMINATION OF THE OPTIMUM AMOUNT OF POLYMER DISPERSANT

The amount of the polymer added to the suspension must be just enough to completely cover the SiC particle surface. The addition of excess may cause an increase in the viscosity and if less than the required amount is added, bridging flocculation may occur and reduce the solids loading. The here used molecular weights of the dispersants were chosen according to their radius of gyration and the van der Waals forces they have to overcome. Rheology was used to decide the optimum concentration of the polymer, i.e. it was assumed that according to above mentioned theories a minimum viscosity must indicate the overall optimized colloidally stable condition [1, 2, 3]. The optimum amount was obtained from the data shown in **figure 1**. Optimum amounts for colloidal stability and viscous behavior were found to be 0.4, 0.3 and 0.4 wt% for polystyrene, polybutadiene and LP1, respectively.

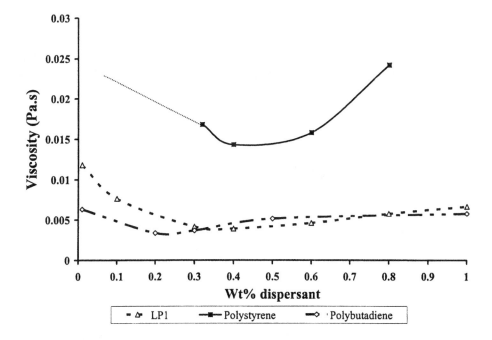

Figure 1: Rheology of 5 vol% SiC, solids loading slurries with variation in dispersants and dispersant concentration. Dispersant amounts are reported as wt% to dry SiC powder. For polystyrene below 0.3wt% (the dotted line) shows the expected curve .

DETERMINATION OF THE OPTIMUM AMOUNT OF CHARGE CONTROL AGENT

Electro-deposition using ESFF is done in non-aqueous dispersing media. This means that the standard charging mechanisms of particles with surface hydroxyls cannot be directly applied. Therefore compounds are added to increase the surface charge of the dispersed particle system. This is typically done with charge controlling agents. Particle charging agents are added to control the magnitude and sign of the charge and for better dispersability of the toner [4]. The optimum amount of the charge controlling agent added to the suspension is taken as the minimum amount that must be added to get the highest amount of uniform deposition (optically checked) on a steel electrode when dipped into the slurry with the application of several kV. Here +4kV were chosen.

Figure 2 presents the optical density of the electrodeposited layer depending on CCA7 for slurries with LP1. The optical density is the ratio of the deposited mass to the surface area on which the deposition was done. The maximum optical density and uniform deposition (optically checked) is seen when the amount of CCA7 is 0.1 times the amount of polymer added. This value is considered as the optimum amount for this set of experiments. For polystyrene, the optimum amount of CCA7 is found to be 0.1 times the amount of the dispersant added.

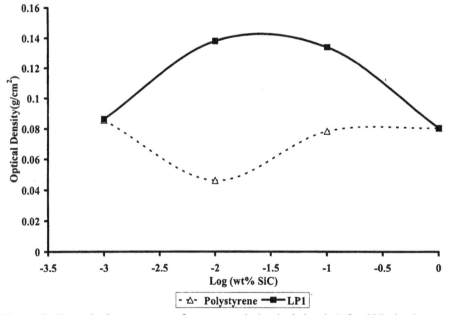

Figure 2: Deposited mass per surface area unit (optical density) for SiC slurries stabilized with polystyrene and LP1.

Colloidal Ceramic Processing

VOLUME FRACTION DEPENDENCE ON RELATIVE VISCOSITY

The deposited layer should be uniform in its properties, i.e. holes, density gradients and deposition of agglomerates have to be avoided. There are several methods that could be used to distinguish the quality of the colloidal dispersion. Since viscosity and reasonably high solids loadings of slurries are also of interest, the Krieger- Dougherty equation is used to evaluate the colloidal quality of the slurries. It can be assumed that if high solids loadings can be achieved the number of agglomerates must be at a minimum. The viscosity of the slurry depends on the maximum solids loading and this value reaches infinity at the maximum volume fraction (Φ_m). For fully dispersed slurry the maximum volume fraction depends on the particle size and the particle shape. At the maximum volume fraction, the particles in the slurry are so close together that their average separation distance is almost zero and this makes their flow impossible [2,3]. **Figure 3** shows the modified Krieger-Dougherty (Equation 1) equation fit to the experimental points. [1].

The best fit of the experimental data gives that Φ_m is drastically lower for the suspension with polystyrene (Φ_m= 0.22 and [η] = 12.8) compared to the LP1 suspension (Φ_m= 0.55 and [η] = 5.41) or the polybutadiene suspension (Φ_m= 0.69 and [η] = 5.24). The lower Φ_m value illustrates that the packing behavior is poor in these cases. The viscosity measurements were done at 298K. The viscosity data was taken at a shear rate of $99.9s^{-1}$.

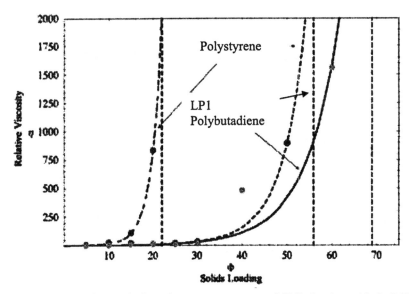

Figure 3: Experimental viscosity measurements of SiC slurries with 3 different dispersants. Curves are fits to equation 1 and the Φ_m values were calculated to be

0.22, 0.55 and 0.69 for the polystyrene, LP1 and polybutadiene suspensions, respectively.

For ESFF applications [5, 6 and 7], adhesion non- adhesion tests were done to find out the best volume fraction to work with. For the adhesion non-adhesion tests, a steel electrode was dipped into slurries with LP1, polybutadiene and polystyrene as dispersants with variation in solids loading (5, 10, 15, 20, 30, 40, 50 and 60 vol% SiC). The aim was to have no particles (SiC or CCA7) sticking to the electrode after dipping for 60 s. A general trend was observed with lower volume fraction slurries performing better than higher solids loadings. Best results with no particles sticking to the electrode were found for polystyrene and LP1 stabilized slurries at 5 vol% SiC. Despite having excellent dispersing abilities polybutadiene always caused particles to be sticking. So, 5 vol% SiC were chosen for further tests. These results will be published elsewhere.

CONCLUSIONS

The optimum amounts for polystyrene, polybutadiene and LP1 dispersants are 0.4, 0.3 and 0.4 wt% for α-SiC [d_{50}=0.52±0.02 μm, 15m^2/g, H.C. Starck]. The optimum amount of charge controlling agent (CCA7) is 0.1 times the amount of polymer added. The maximum solids loading attainable calculated using the Krieger-Dougherty fit equation for slurries with polystyrene was 0.22 [η=12.8], with LP1 was 0.55 [η=5.41] and with polybutadiene was 0.69 [η=5.24]. It can be seen that the slurries with polystyrene or LP1 as the polymer satisfy the initial adhesion non-adhesion test and its seen that 5 vol% SiC is the best volume fraction to work with for the rapid prototyping applications.

REFERENCES

[1]Y.Yang, W.Sigmund,(2001), "Preparation, Characterization, and Gelation of Temperature Induced Forming (TIF) alumina slurries", Journal of materials synthesis and processing, Vol 9, No 2
[2]M. N.Rahaman,(1995), "Ceramic Processing and Sintering", Marcel Dekker Inc
[3]W.Sigmund, N.Bell, L.Bergstrom,(2000), "Novel Powder Processing Methods For Advanced Ceramics", Journal of American ceramics society, 83[7]1557-74.
[4]Prof J.Thomas, University of Southern Australia, "Developing a Dielectric Spectrometer" http://www.laser.unisa.edu.anonpolar.htm, Particle charging in non polar media
[5]V. Kumar, (2000), "Solid Freeform Fabricate on Using Powder Deposition", US Patent 6,066,285.
[6]V.Kumar, S. Rajagopalan, M.Cutkoshy and D.Dutta, (Dec 7-9, 1998), "Representation and Processing of Heterogenous Objects for Solid Freeform Fabrication", IFIP WG5.2 Geometric modeling Workshop,Tokyo.

[7]L.Schein,(1992),"Electrophotography and Development Physics", Springer series in electrophysics 14; Springer Verlag.

ANALYSIS OF ACTION MECHANISM OF ANIONIC POLYMER DISPERSANT WITH DIFFERENT MOLECULAR STRUCTURE IN DENSE SILICON NITRIDE SUSPENSION BY USING COLLOIDAL PROBE AFM

Hidehiro Kamiya, Shinsuke Matsui
Graduate School of Bio-Applications and Systems Engineering, BASE
Tokyo University of Agriculture & Technology, Koganei, Tokyo 184-8588, Japan

Toshio Kakui
LION Co., Ltd., Tokyo 132-0035, Japan

ABSTRACT

The present paper focused on the action mechanism of anionic polymer dispersants in dense Si_3N_4 suspension with oxide additives. New preparation method of colloidal probe in which spherical single particle with several μ m in diameter was adhered on the top of a commercial tip for AFM, was developed by using micro-manipulation system for bio-technology. Copolymer of ammonium acrylate and methyl acrylate with different hydrophilic to hydrophobic group ratios, m:n, was prepared and added to the suspension. Microscopic surface interaction between Si_3N_4 and Si_3N_4 or oxide in solution was characterized by using colloidal probe atomic force microscopy (AFM) and model anionic polymer dispersants. The relationships between macroscopic suspension viscosity and the microscopic surface interaction were analyzed. Based on these results, the adsorbed mechanisms of polymer dispersant on silicon nitride and oxide particles and hetero-coagulation phenomena in suspension were discussed.

INTRODUCTION

Hard agglomerates limit uniform consolidation of green compacts for ceramics and leave large pores in the resultant sintered bodies of silicon nitride ceramics[1, 2]. Various kinds of polymer dispersant have been used to control aggregation and dispersion behavior of fine particles in suspension and various mechanisms of polymer dispersant in suspensions (steric or overlapping force, re-bridging and depletion, etc.) were discussed by way of explaining the suspension behaviors of different additive conditions. However, the optimum molecular weight[3, 4] and additive conditions of a dispersant[5, 6] were determined based on empirical data gathered during macroscopic suspension behavior, for example, suspension viscosity and aggregate size distribution. However, these methods do not provide direct information about the interaction among particles with polymer dispersant in dense suspensions.

The surface on silicon nitride powders in aqueous suspension system has a complex molecular structure. Hydroxyl group and ammonium group have been formed on the surface in the reaction with the water. Anionic polymer dispersants are used to reduce the viscosity of suspension with silicon nitride.

In our previous work[7], ammonium polyacrylate with different hydrophilic to hydrophobic ratios (m:n), was prepared and added to dense alumina suspensions (40 vol%). The steric interactions and adsorbed structures of dispersants on alumina powders were examined under an atomic force microscope (AFM). An optimum hydrophilic to hydrophobic group ratio, which was obtained from the maximum repulsive force and the minimum viscosity of suspension, was determined at m:n = 3:7 as shown in Fig. 1. The molecular structure design of polymer dispersant is important to control of aggregation and dispersion behavior of ceramic powder in dense aqueous suspension.

This paper focuses on the action mechanism of anionic polymer dispersant in silicon nitride dense suspension. The effect of polymer dispersant adsorption on the macroscopic suspension behavior and microscopic interaction between silicon nitride surfaces was analyzed by using copolymers of methyl acrylate and ammonium acrylate with different hydrophilic to hydrophobic ratio[5] and colloidal probe AFM[6]. Based on the comparison of microscopic interaction and macroscopic suspension viscosity, the action mechanism of anionic polymer dispersant in silicon nitride suspension is discussed.

Colloidal Ceramic Processing

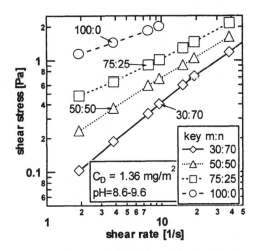

Fig. 1 Shear stress and rate relationship of dense alumina suspension with different molecular structure polymer dispersant. Effect of hydrophilic to hydrophobic group ratio[7]

EXPERIMENTAL PROCEDURE

(1) Preparation and characterization of suspension

A high purity silicon nitride (SN-E10, Ube-Kousan Co.Ltd., Japan) was used as powder raw material. Polymer dispersant used was ammonium polyacrylate with different hydrophilic to hydrophobic ratio (m:n = 30:70 and 100:0) with the same average molecular weight of 10,000. The silicon nitride powder was mixed with water containing different concentrations of polymer dispersant, and ball-milled for 24 hours. The solid volume fraction of silicon nitride in the suspensions was set at 35 vol%. The additive content of the polymer dispersant in solution before ball milling ranged from 0.01 to 2 mg/m^2. Suspension viscosity was determined by a concentric cylinder viscometer at a shear rate ranging from 1 to 1000 s^{-1}.

(2) Preparation of colloidal probe for an AFM

New preparation system and method of colloidal probe, in which spherical single particle with several μm in diameter was adhered on the top of a commercial tip for AFM, was developed by using micro-manipulation system for bio-technology. The schematic of this equipment is shown in Fig. 2. A commercial tip of AFM was connected with the micro-manipulator. Polymer binder was applied on the surface of needle 1, and many spherical particles were adhered on the surface of the needle 2. Firstly, in order to apply the polymer binder to the tip of a commercial tip, the top of tip was contacted the surface of binder on the needle 1. Secondly, the tip of AFM was approached and picked up one particle on needle 2.

Since the mean diameter of silicon nitride powder is about 0.1 μ m, it is impossible to prepare colloidal probe from single particles. Spray-dried granules with several μ m in diameter were prepared from silicon nitride suspension, and sintered at 1373 K in nitrogen. By such a relatively low temperature heat treatment, the neck growth between primary silicon nitride particles was generated, and the strength of granule increased. By using spray-dried granule after heat treatment, an example of a colloidal probe prepared by the above system was shown in Fig. 2.

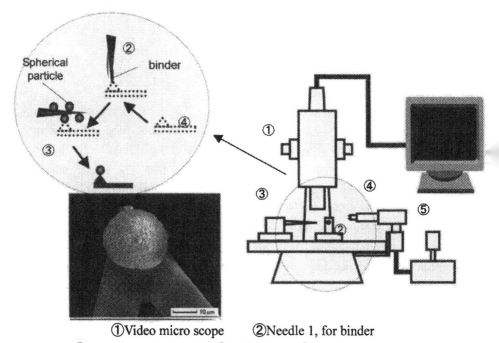

①Video micro scope   ②Needle 1, for binder
③Needle 2, for particle   ④AFM probe ⑤Micro manipulator
Fig. 2 Schematic of preparation system of a colloid probe for AFM and example of silicon nitride probe used spray-dried granule.

Colloidal Ceramic Processing

(3) Measurement of electrosteric interaction using a colloidal probe AFM

Variation of the electrosteric repulsion force on the polished surface of porous silicon nitride in solution with different polymer dispersant was examined by AFM with a colloidal probe method. The interaction between a colloidal probe of silicon nitride and a polished surface of a sintered silicon nitride piece was measured. Sintered silicon nitride pieces and spherical probe were prepared from same silicon nitride powder materials.

In order to adsorb the polymer dispersant onto a polished piece (under the same condition of fine powders in suspension), the polished sintered silicon nitride piece was mixed into a dense silicon nitride. After ball milling of the silicon nitride slurry with a polished sintered silicon nitride piece for 24 hours, the dense powder layer with the thin piece and the clear layer were separated by the centrifugal sedimentation method. The sintered piece was picked up from the dense powder layer and washed by the clear layer, and transferred into the AFM fluid cell. The electrosteric repulsion force and adhesion force were determined as distance approached 0 and departed from the contact condition in the clear layer, respectively.

RESULTS

(1) Suspension viscosity of dense silicon nitride suspension

Figure 3 shows the effect of molecular structure and additive content of polymer dispersant on apparent viscosity of dense silicon nitride suspension. The solid fraction of silicon nitride powder in all suspensions was 35 vol%. The hydrophilic to hydrophobic ratio of polymer dispersants used are 30:70 and 100:0.

The optimum additive content of polymer dispersant to obtain the minimum viscosity is about 0.01 mg/m^2. When the polymer dispersant with 100 % hydrophilic group was used, the lower value of suspension viscosity was observed. To analyze these effects of molecular structure of polymer dispersant on silicon nitride suspension viscosity, the surface interactions between silicon nitride were determined by using a colloidal probe AFM.

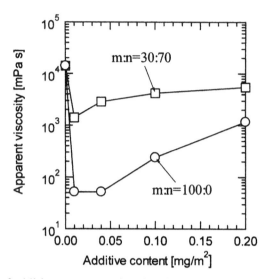

Fig. 3 Effect of additive content and molecular structure of polymer dispersant on apparent suspension viscosity, shear rate = 10 s^{-1}.

(2) Surface interaction between silicon nitride

Figure 4 shows the effects of molecular structure of polymer dispersant on the force curve between a silicon nitride colloidal probe and polished silicon nitride substrate. The force curve determined by DLVO theory is shown in this figure. A theoretical curve based on DLVO theory almost agreed with the experimental results at the relatively long surface distance > 8 nm. If the additional steric forces do not act at the solid-water interface, van der Waals type attractive forces appear at a distance of about 3 - 5 nm. However, the repulsive interaction of the experimental force curve with both dispersant continued to increase at a short distance of < 8 – 10 nm. If each polymer dispersant was added in solution, the additional short ranged repulsive force was observed at a relatively short range below about 10 nm by comparing with the theoretical curve. The difference between the experimental force curve and DLVO theory at this short distance was steric repulsive force by the loop - train structure formation of adsorbed polymer dispersant. This short-range repulsive force clearly depends on the hydrophilic to hydrophobic ratio of the dispersant.

It is assumed that this short-range repulsive force is the steric interaction of adsorbed polymer on a solid surface. The steric repulsive force in solution with

Colloidal Ceramic Processing

polymer dispersant of 100 % hydrophilic group was larger than that with 30 % hydrophilic group. The shortage of hydrophilic group in the dispersants changed the adsorbed structure of dispersants on solid surfaces and hindered the development of steric forces.

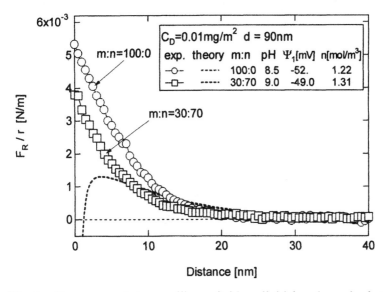

Fig. 4   Force curve between silicon nitride colloidal probe and substrate.

Figure 5 shows the adhesion strength distribution between silicon nitride colloidal probe and substrate. Many data of the adhesion force were measured at different position on silicon nitride substrate. Based on the measurement values on the different position on the substrate, the adhesion force distribution was calculated and shown in Fig. 5. Adhesion strength between silicon nitride colloidal probe and substrate has a wide distribution in solution without polymer dispersant. The mean value of adhesion force was about 12 µN/m. When polymer dispersant was added in suspension, both polymer dispersants adsorbed on silicon nitride and reduced the attractive force and disappeared at some place on substrate. Since the adsorbed polymer dispersant protected the direct contact of solid surface, the attractive force was reduced and disappeared. The increase of short-range steric repulsive force and the decrease of adhesion attractive force by the adsorption of polymer dispersant prompted the dispersion of aggregates in suspension and decreased the suspension viscosity.

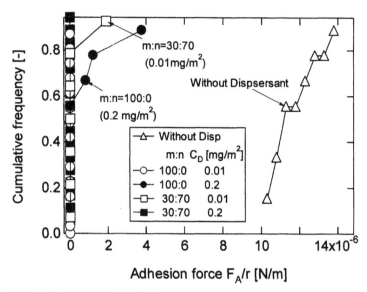

Fig. 5   Adhesive strength distribution between silicon nitride colloidal probe and substrate in solution with different additive content and molecular structure of polymer dispersant.

DISCUSSION

Based on the above results, the effects of molecular structure on adsorbed structures on solid surface are shown in Fig. 6. It seems that the hydrophilic group in polymer dispersant adsorbed at neutral charged hydrophilic group ($-NH_2$) on solid surfaces. Since the density of this $-NH_2$ group is not so high, the number of adsorbed points of dispersants on solid surface seems to be very little. When the hydrophobic group increased (for example, m:n =30:70), the number of adsorbed points of dispersant on solid surfaces was not sufficient to form a loop or train structure as shown in Fig. 6 (a). Since a few hydrophilic group of the dispersant can be only adsorbed as a tail structure on the silicon nitride surface, the steric force of polymer dispersants of low hydrophilic group ratio (m:n = 30:70) decreased.

On the contrary, when high hydrophilic group polymer dispersant is used, since multiple adsorption points per one dispersant on silicon nitride surface exist, polymer dispersant with 100% hydrophilic group can be formed loop and train structure, which accelerated a steric effect on the silicon nitride surfaces. These adsorbed structures accelerated the steric force of the dispersants and reduced the adhesion force between solid surfaces. High steric repulsive force and low adhesive interaction between silicon nitride surfaces promoted the particle dispersion and reduced suspension viscosity.

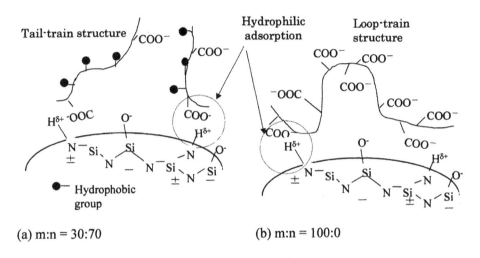

(a) m:n = 30:70   (b) m:n = 100:0

Fig. 6 Estimated adsorbed structure of polymer dispersant.

CONCLUSION

The optimum molecular structure of anionic polymer dispersant in Si_3N_4 suspension to obtain high electro-steric repulsion force between silicon nitride and low suspension viscosity were determined at the hydrophilic to hydrophobic ratio, m:n = 100:0. The effect of hydrophilic to hydrophobic ratio on adsorbed structure on silicon nitride surface in aqueous suspension was estimated based on the surface interaction characterized by a colloidal probe AFM method.

ACKNOWLEDGEMENT

This study was supported by a Grant-in-Aid for Scientific Research (B) from the Japanese Ministry of Education, Science, Sports and Technology, and the

Structurization of Material Technology Project in the Nano-Technology Program by METI Japan.

REFERENCES

1. H. Kamiya, K. Isomura, G. Jimbo and J. Tsubaki, J. Am. Ceram. Soc., 78(1), 46-57 (1995)
2. H. Kamiya, M. Naito, T. Hotta, K. Isomura, J. Tsubaki and K. Uematsu, Am. Ceram. Soc. Bull., 76(10), 79 (1997)
3. V. A. Hackley, J. Am. Ceram. Soc., 80(9), 2315-25 (1997)
4. H. Okamoto, M. Hashiba, Y. Nurishi, K. Hiramatsu, J. Mater. Sci., 26, 383-387 (1991)
5. J. Cesarano III and I. A. Aksay, J. Am. Ceram. Soc., 71(12), 1062-67 (1988)
6. H. Unuma, B.H. Ryu, I. Hatano and M. Takahashi, J. Soc. Powder Technol. Japan, 35(1), 25-30 (1998)
7. H. Kamiya, Y. Fukuda, Y. Suzuki, M. Tsukada, T. Kakui and M. Naito, J. Am. Ceram. Soc., 82(12), 3407-12 (1999)
8. W. A. Ducker and T. J. Senden, Langmuir, 8, 1831-36 (1992)

AQUEOUS PROCESSING OF WC-Co POWDERS: SUSPENSION PREPARATION AND GRANULE PROPERTIES

Karin M Andersson and Lennart Bergström
YKI, Institute for Surface Chemistry
PO Box 5607
SE-114 86 Stockholm
Sweden

ABSTRACT

This paper focuses on how a thorough understanding of the surface chemistry of complex systems can assist in the design of a water-based process. We have investigated different aspects of the powder processing, including slurry preparation and granule characterisation. We have followed the oxidation and dissolution kinetics of WC and WC-Co mixtures, and found that the presence of Co suppresses the dissolution of WC. We have used the atomic force microscope (AFM) colloidal probe technique to measure the surface forces resulting in electrosteric stabilisation of WC-Co suspension with poly(ethylene imine) (PEI). The AFM was also used to measure the friction forces between individual spray-dried granules and a pressing tool wall. It was found that friction is largely affected by binder content and relative humidity. Moreover, a new technique to measure the density of spray-dried granules is presented.

INTRODUCTION

Hard metals are commonly produced following a powder metallurgy route, which involves powder production, mixing, granulation, pressing and sintering[1]. Each of these manufacturing steps must be optimised to yield a dense product with a homogeneous microstructure. The reliability and strength are controlled by the abundance and size of the defects, which can have their origin in the powder (hard agglomerates, large grains), processing (organic inclusions, density gradients, cracks), or sintering (abnormal grain growth, cracks)[2]. Agglomerates in the hard metal powder are supposedly broken down during the wet milling step. Traditionally, organic solvents have been used as grinding fluid, but for environmental reasons, the use of water is encouraged. However, handling non-oxide powders in aqueous media can lead to oxidation of the powders and problems with reprecipitation of dissolved material. Hence, the solubility and the dissolution rate of the dispersed powders can have a significant influence on the final properties.

Although the milling procedure may be efficient in breaking down the inherent agglomerates, care has to be taken to prohibit the formation of detrimental agglomerates at a later stage, e.g. during spray drying. A sufficient colloidal stability, governed by the interparticle forces, is therefore required. One way of obtaining a well-dispersed suspension is to introduce a repulsive steric force. Polymers and polyelectrolytes are much-used dispersants for various powders suspended in a liquid phase[3-4]. Laarz and Bergström successfully dispersed WC-Co slurries with a high powder fraction using poly(ethylene imine) (PEI)[5], which is a highly charged poly-electrolyte.

After dispersing and milling the slurry, the fine powder needs to be transformed into larger granules in order to make the powder free flowing and suitable for automatic feeding. This is commonly done by spray-drying. A polymeric binder is added to the suspension prior to spray-drying to impart a sufficient strength to the granules. The amount of binder will largely affect granule properties such as granule strength, density, and friction behaviour and thus the pressing of the powder body[6-10].

This paper focuses on the behaviour of the WC-Co powder in aqueous media, including dissolution, interparticle forces, and granule properties. The kinetics of the dissolution process of the WC and WC-Co powder mixtures has been investigated. The atomic force microscope (AFM) colloidal probe technique, has been used for investigating the electrosteric forces caused by the adsorption of PEI to WO_3 surfaces. Moreover, new techniques for measuring the properties of spray-dried granules from aqueous suspensions with varying amount of poly(ethylene glycol) (PEG) binder are presented. The friction between a single spray-dried granule against a polished hard metal surface is measured as a function of load with the AFM colloidal technique. The density of individual granules has also been measured using the AFM.

EXPERIMENTAL

The WC powder had a BET specific surface area of 0.49 m^2 g^{-1}. The BET specific surface area of the Co powder was 1.15 m^2 g^{-1}. X-ray diffraction analysis (XRD) showed that the Co powder consists of a mixture of the cubic and the hexagonal phase. XPS measurements were carried out on WC and WC-Co powder mixtures in order to deduce the degree of oxidation. X-ray photoelectron spectroscopy (XPS) spectra were recorded using a Kratos AXIS HS X-ray photoelectron spectrometer, with a Mg K α X-ray source operated at 240 W (12 kV / 20 mA).

The dissolution of commercial WC and WC-Co powder mixures was investigated at different pH-values by continuously taking out samples from 2 vol% aqueous suspensions and measuring the concentration of Co and W in solution using inductively coupled plasma (ICP) analysis. The suspensions were kept in closed plastic containers and shaken. The pH was kept at set values between pH 3 and 11 and regularly checked and adjusted by additions of acid or base. Samples for analysis were obtained by separating a clear supernatant from the powder using filtration.

Surface force measurements were performed with an atomic force microscope (AFM) using the colloidal probe technique[11]. The normal surface forces between the colloidal probes and the flat surfaces were evaluated by compressing and retracting the surfaces with a piezo electric crystal scanner. The deflection of the cantilever was registered as a function of the movement of the scanner in the Z direction using a laser beam, reflected off the cantilever onto a photodiode detector. The spring constant of the cantilever was used to transform the deflection data into a force, F, through Hooke's law

$$F = kD \qquad (1)$$

where k is the spring constant of the cantilever and D is the deflection of the cantilever. The spring constants of the cantilevers were calibrated using the added mass technique[12] and were in the range 0.10-0.12 N/m.

Oxidised tungsten metal micro-spheres with a radius in the range 5.4-7.0 μm were used as colloidal probes. XPS analysis of the micro-spheres and a commercial WC powder shows that both materials are covered with a thin WO_3 layer, which indicates that we are using a system of technical relevance. A flat polished cobalt metal surface was used as a substrate. XRD analysis of the oxidised cobalt surface showed diffraction peaks that matched literature data of the CoOOH phase.

The AFM measurements were carried out in aqueous electrolyte solutions using a sealed liquid cell. All electrolytes were pH adjusted with additions of diluted NaOH solutions before they were injected into the cell. The poly(ethylene imine) (PEI) from Aldrich Chemical Company Inc., USA, was branched with a weight-averaged molecular weight, M_w, of 25 000. Force measurements were performed between surfaces that had been exposed to 500 ppm PEI solutions. The pH was 9.0 and the PEI was left to adsorb for 2 hours.

The AFM colloidal probe technique has previously been shown to be a powerful tool to measure the friction between a granule and a flat surface and also between individual granules[13]. Spray-dried WC-Co (10wt% Co) granules are glued onto the apex of rectangular cantilevers and used as colloidal probes. Friction measurements are attainable by recording the torsional bending of the cantilever as it slides sidewards a specific distance across the surface. The signal is readily available from the detector and can subsequently be converted into friction at a particular applied load. The load is increased in steps and then increased to obtain loading-unloading curves. The granules were produced by spray-drying aqueous WC-Co (10wt% Co) suspensions containing 0.5wt% and

2wt% PEG 4000. The radii of the granules used as probes were typically 10 μm. The friction is measured against a polished surface of a hard metal pressing tool.

DISSOLUTION OF WC-Co POWDERS

XPS measurements show that the as-received industrial WC powder has an oxidised surface. We have previously shown that the ratio W_{WO3}/W_{tot} decreases with an asymptotic approach to a constant value at long immersion times[14]. Hence, the thickness of the oxide layer on the WC powder decreases when exposed to aqueous media; the oxide layer is slowly dissolved until a steady-state is reached, where the dissolution and re-oxidation rates are equal. The W 4f spectrum contains contributions from both WC and WO_3 at all immersion times, thus the WC powder is always covered with a thin WO_3 layer in aqueous suspension. WO_3 dissolves in water forming tungstate ions and protons by the reaction:

$$WO_3 + H_2O \leftrightarrow WO_4^{2-} + 2H^+ \tag{2}$$

At low pH, W(VI) forms a number of different polynuclear species in solution[15-19].

We find that the dissolution rate of the oxidised WC in water is somewhat higher at high pH due to the acidity of the oxide. However, the difference between the highest (pH 11) and lowest pH (pH 3) is not significant. The straight line in figure 1a shows the average dissolution rate in WC suspensions with pH values ranging from 3-11.

Co powder also oxidises continuously in water. Contrary to the acidic tungsten oxide, cobalt oxides are basic and dissolve under the formation of hydroxide ions by reactions such as:

$$CoOOH + H_2O \leftrightarrow Co^{3+} + 3OH^- \tag{3}$$

In a mixed WC-Co suspension the release of W decreases (fig. 1a), especially at low pH. Figure 1b shows the dissolution of Co in the mixed WC-Co suspension. We find that Co is more soluble at low pH. The observed reduction in solubility of WC in a mixed WC-Co suspension below pH 11 corresponds to the onset of Co dissolution. This suggests that a Co-W compound, such as cobalt tungstate, $CoWO_4$, with a low solubility product is formed and suppresses the dissolution of WC. The change of pH in a WC-Co powder mixture suspension that is not pH adjusted is relatively small (not shown) and stays between 8 and 9, which shows that the simultaneous dissolution of the oxidised WC and Co powders buffers the suspension.

Colloidal Ceramic Processing

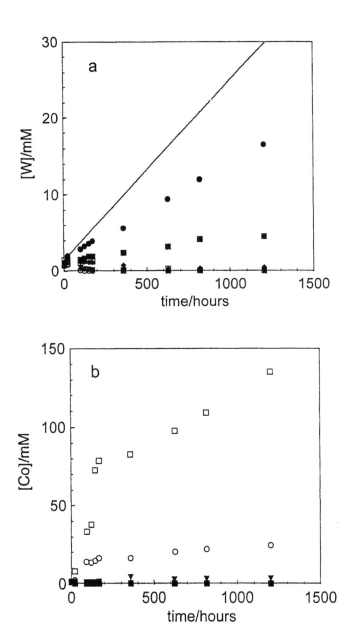

Fig.1 Release of W and Co in a WC-Co (5wt% Co) aqueous slurry with time at different pH-values; pH 3 (□), pH 5 (○), pH 7 (▼), pH 8 (▲), pH 9 (◆), pH 10 (■), pH 11 (●). a) Release of W. The straight line corresponds to the average dissolution rate of W in a WC aqueous slurry at pH 3-11. b) Release of Co.

Colloidal Ceramic Processing   97

SURFACE FORCES: DISPERSION OF WC-Co WITH PEI

The colloidal stability of an aqueous suspension is controlled by the surface forces between the individual particles. The surface forces are governed by the material characteristics and the surface and solution chemistry. A high surface charge of the particles can be enough to stabilise a suspension electrostatically; with a pH dependent surface charge it may be possible to induce a strong repulsive force between particles by altering pH in the suspension. Often this means working at a very high or low pH values, which can cause problems with dissolution and corrosion of process equipment. In order to avoid this, dispersants are commonly used to stabilise a suspension. Dispersants are polymers that adsorb at the solid liquid interface and infer a repulsive force, eliminating adhesion of particles. If the polymer is charged it can give rise to both an electrostatic contribution to the stabilising force at large separations and a steric force at smaller particle separations.

The atomic force microscope (AFM) colloidal probe technique, has been used for investigating the electrosteric forces caused by the adsorption of poly(ethylene imine) PEI to oxidised W and Co surfaces. This technique was originally developed by Ducker et al[11], and is now widely applied. The colloidal probe technique facilitates the investigation of surface forces for a wide range of materials.

The surface forces between the oxidised tungsten and cobalt surfaces as a function of separation in pure NaCl electrolyte are shown in figure 2a. We find that the forces are attractive and that the range of the attraction increases with decreasing ionic strength, which proves that the attraction is electrostatic in origin. This suggests that oxidised tungsten and cobalt surfaces have opposite surface charges at pH 10. We know from previous work that oxidised WC is negatively charged over a wide pH range[14], thus, under these preparation conditions, the oxidised cobalt sample must be positively charged and have an iep higher than 10.

Figure 2b shows the force between the surfaces after being exposed to PEI solution. Exchanging the PEI solution for pure NaCl electrolyte solution simplifies the interpretation of the force curves, as the ionic strength can easily be estimated. From the measurements, two different regions of the compression curves can be identified. The long-distance tail of the curves obtained in 0.1 and 1 mM is obviously of electrostatic origin; the range of the measured forces follows the expected ionic strength dependence (see inset in figure 2b).

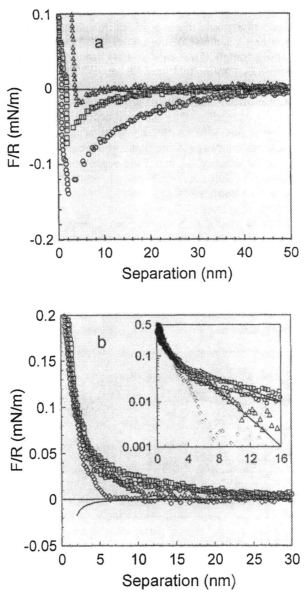

Fig 2 Normalised forces between WO_3 and CoOOH a) in NaCl solutions at pH 10; 0.1 mM (O), 1 mM (□), and 5 mM (△) from ref 20. b) in 500 ppm PEI solution rinsed with NaCl solutions: 0.1 mM (O), 1 mM (□), 10 mM (△), 100mM (◇) from ref 21. The inset shows the curves on a logarithmic scale, and the solid lines are fits to the DLVO-theory, using the constant charge boundary condition.

At closer separations, there is an additional contribution to the force curves. This repulsion is induced by the adsorbed PEI and is most clearly seen at the highest ionic strength (100 mM NaCl), where the magnitude of the electrostatic contribution to the total force, is small. Surface force measurements with the AFM colloidal probe technique has the limitation that the determination of surface separation is not unambiguously presented by the equipment, but is a somewhat subjective estimation of the hard wall contact. Thus, the exact thickness of the adsorbed layer cannot be measured. We do achieve, however, an estimate of the thickness from the range of the repulsive force, which suggests polymer layer thickness in the order of 4-5 nm. This agrees rather well with the quoted radius of the polymer coils in solution, bearing in mind that the coils experience a certain collapse upon adsorption[22-24].

In a symmetric system (WO_3-WO_3), the polymeric interaction starts at a separation distance of around 9nm (not shown). This is approximately twice the separation distance measured in the asymmetric system, suggesting that the PEI adsorbs only to one surface (see figure 3). Assuming that the positively charged PEI again adsorbs to the negative WO_3 surface, this adsorbed layer interacts with the positively charged bare oxidised cobalt surface, giving rise to an electrostatic repulsion. Hence, adsorption of the positively charged PEI to the WO_3 surface in this system, results in a repulsive electrosteric force between the WO_3 surface and the oxidised cobalt surface.

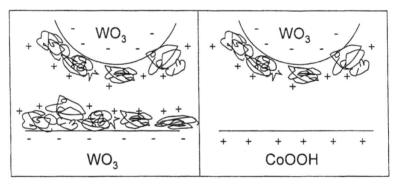

Fig 3. Schematic drawing of the origin of the repulsive forces between WO_3 surfaces (left), and WO_3 and CoOOH surfaces (right) in a PEI solution.

Colloidal Ceramic Processing

GRANULE FRICTION

Industrial compaction of fine powders requires granulation in order to facilitate handling and feeding prior to die filling. Granulation is commonly achieved by spray-drying of a powder suspension that has been optimised with respect to colloidal stability, solids loading and binders[7, 23-26]. A binder commonly used for powder pressing is poly(ethylene glycol) (PEG). In the spray-drying process, the water-soluble binder is accumulated at the granule surface[27-28]. Therefore, the properties of the binder will influence the friction behaviour of the granule in the compaction step. Significant friction at the tool wall during pressing may yield low-density areas of incomplete granule deformation within the compact. We have used the AFM to measure the friction as a function of applied load between single granules and a pressing tool surface.

Figure 4 a shows the friction force as a function of load between a spray-dried WC-Co granule containing 0.5 wt% PEG and a hard metal surface at three different relative humidities. The friction is relatively low and the dependence of humidity on the friction force is small. We also find that the adhesion force, defined as the friction force at zero applied load, is small.

The friction and adhesion increases when the polymeric binder content is increased, especially at high relative humidity (fig 4b). The friction between the hard metal surface and granule is obviously affected by the properties of the PEG binder. The hygroscopic nature of PEG makes the binder sensitive to changes in relative humidity. The water uptake softens the PEG by lowering its shear modulus, thereby affecting adhesion and friction. When the granule surface is hard and has limited ability to deform under pressure, the friction forces are mainly governed by surface morphology. However, when the binder-rich, surface is slid against the substrate, binder softening causes viscous surface deformation. We find that at 70% relative humidity there is a strong hysteresis between the loading and unloading curves. This is a result of the water uptake that makes the binder-rich granule surface soft and deformable. The granule deforms slightly upon loading, which increases the contact area. When the applied load is decreased the contact area remains high and results thus in an increased friction force during unloading.

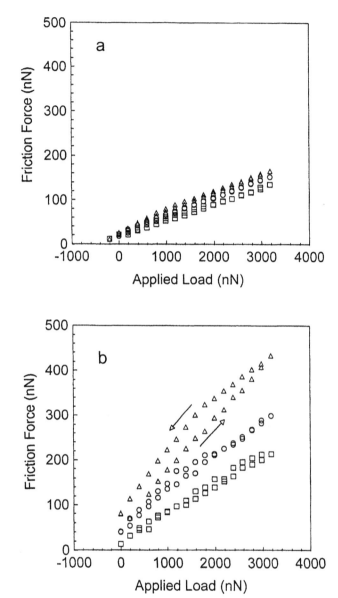

Fig. 4. Friction-load measurements between spray-dried WC-Co (10wt%Co) granule and a flat hard metal surface. a) Loading-unloading friction curves for a granule containing 0.5wt% PEG at relative humidity of; 5% (□), 67% (O), and 79% (△). b) Loading-unloading friction curves for a granule containing 2wt% PEG at relative humidity of; 5% (□), 64% (O), and 70% (△).

Colloidal Ceramic Processing

GRANULE DENSITY

The density of the spray-dried granules is an important property affecting the compaction behaviour of the green body[10, 28]. On the one hand a high porosity of the granules may result in a low-density body after compaction. On the other hand, a high density may increase the strength of spray-dried granules leading to incomplete deformation and intergranular pores.

Here we have used the AFM as a sensitive balance to determine the mass of individual granules. An AFM cantilever is a beam type Hookean spring and the fundamental resonance frequency, v_0, is described by the solution of the equation of motion of the lever which gives:

$$v_0 = \frac{1}{2\pi}\sqrt{\frac{k}{m_{eff}}}$$

(4)

where m_{eff} is an effective mass depending on the cantilever. When a cantilever is loaded by a mass, M, (e.g. a particle) at the end the loaded resonance frequency, v_l is

$$v_l = \frac{1}{2\pi}\sqrt{\frac{k}{m_{eff} + M}}$$

(5)

Combining equations (4) and (5) gives:

$$k = \frac{(2\pi)^2 M}{(1/v_l^2) - (1/v_0^2)}$$

(6)

Using this relation, the spring constant, k, of a cantilever can be determined when spherical particles of known mass are added to the cantilever and the resonance frequencies, v_0 and v_l are measured[12].

In this study we have used cantilevers with known spring constants to determine the mass of spray-dried granules. Using a micromanipulator granules with radii in the range 5-40 µm were placed at the end of cantilevers with k=1.3-75 N/m. The fundamental and loaded resonance frequencies were measured in the AFM. The granules were produced by spray-drying aqueous WC-Co (10wt%Co) suspensions containing 0.5wt% and 2wt% PEG 4000. The volume of the granules was determined by measuring the radius with an optical microscope.

The theoretical compact density of WC-Co (10wt%Co) containing 0.5wt% and 2wt% PEG is approximately 13.7 and 11.2 g cm^{-3}, respectively. The measured granule densities were on average 6.29 g cm^{-3} for granules with 0.5wt% PEG and 6.06 g cm^{-3} for granules with 2wt% PEG over the whole range of measured radii. The porosity can be expressed as the fraction of the granule volume being pores. Figure 5 shows the porosity of spray-dried granules containing 0.5wt% and 2wt% PEG. The average porosity of the granules containing 2 wt% PEG is 0.46. For the granules with less PEG binder (0.5wt%) the porosity is somewhat higher, 0.54 on average. These results show that the polymeric binder contributes to a higher compactness of the spray-dried granules.

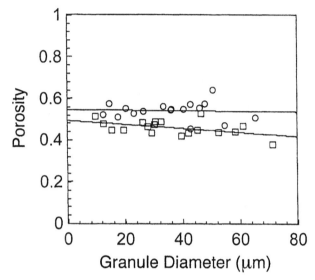

Fig. 5 Porosity of spray-dried WC-Co (10wt%Co) granules containing 0.5 wt% PEG 4000 (O) and 2wt% PEG 4000 (□).

Colloidal Ceramic Processing

SUMMARY AND CONCLUSIONS

We have described various aspects of the aqueous processing of oxidised WC-Co powders and how the AFM could be used for characterisation of individual WC-Co granules. It was shown that the solubilities of WC and Co powder have opposite pH dependence. The dissolution rate of WC powder in aqueous suspension decreases in the presence of Co powder, which is directly related to the cobalt concentration in the liquid phase. We speculate that the formation of a Co-W compound such as $CoWO_4$ may be responsible for suppressing the dissolution rates of both powders, keeping the pH at a user friendly value of around 8-9.

We have evaluated the interactions between oxidised tungsten and cobalt surfaces in poly(ethylene imine) (PEI) solution using the atomic force microscope (AFM). We found that PEI adsorbs on WO_3 and induces a net repulsive interaction in the asymmetric system WO_3-CoOOH. The conversion from an attractive interaction, before addition of PEI, to a repulsive, is the basis of the use of PEI as a dispersant in hard metal suspensions. The steric part of the repulsive force is more short-range than in the symmetric system, which probably can be explained by adsorption of PEI only to the WO_3 surface and not on the oxidised Co.

We have used the AFM to measure the friction as a function of applied load between spray-dried WC-Co granules and a pressing tool surface. The dependence of relative humidity on the friction was shown to increase with increasing amount of poly(ethylene glycol) (PEG) binder in the granule. The results clearly illustrate the effect of binder concentration and relative humidity on the friction and adhesion between granules in a powder body during die pressing.

We have developed a new method where the AFM is used to determine the density of spray-dried granules. An AFM cantilever is used as a balance for mass determination of individual granules. The change of the resonance frequency when a granule is placed on the cantilever is used to calculate the mass of the granule. We have measured the density of spray-dried WC-Co (10wt%Co) granules containing 0.5wt% and 2wt% PEG 4000 binder. It was found that the granules containing 2wt% PEG had an average porosity of 0.46 whereas the average porosity for the 0.5wt% PEG granules was 0.54.

ACKNOWLEDGEMENTS

This work has been performed within the Brinell Centre Inorganic Interfacial Engineering (BRIIE), supported by the Swedish Agency for Innovation Systems (VINNOVA) and the following industrial partners: Erasteel Kloster AB, Höganäs AB, AB Sandvik Coromant, Seco Tools AB and Atlas Copco Secoroc AB.

Marie Ernstsson is thanked for XPS measurements and interpretation; Mats Johnsson at Stockholm university for XRD and BET surface area measurements. Eric Laarz is acknowledged for helpful comments. Olivier Alvain (CERMeP, Grenoble) is thanked for providing the spray-dried granules.

REFERENCES

[1]R.M. German, "Powder Metallurgy Science", 2nd ed. Metal Powder Industries Federation, Princeton, 1994.

[2]D. Sherman and D. Brandon, "Mechanical Properties and Their Relation to Microstructure"; pp. 66-103 in *Handbook of Ceramic Hard Materials Vol 1*, 1st ed. Edited by R. Riedel, Wiley, Weinheim, 2000.

[3]R.J. Pugh, "Dispersion and Stability of Ceramic Powders in Liquids"; pp. 127-192 in *Surface and Colloid Chemistry in Advanced Ceramics Processing*, 1st ed. Edited by R.J. Pugh, L. Bergström, Marcel Dekker, New York, 1994.

[4]D.J. Shanefield, "Organic Additives and Ceramic Processing", 2nd ed. Kluwer Academic Publishers, Boston, 1996.

[5]E. Laarz and L. Bergström, "Dispersing WC-Co Powders in Aqueous Media with Polyethyleneimine," *International Journal of Refractory Metals and Hard Materials,* **18** [6] 281-286 (2000).

[6]H. Kamiya, K. Isomura and G. Jimbo, "Powder Processing for the Fabrication of Si_3N_4 Ceramics: I, Influence of Spray-Dried Granule Strength on Pore Size Distribution in Green Compacts," *Journal of the American Ceramic Society,* **78** [1] 49-57 (1995).

[7]S.J. Lukasiewicz and J.S. Reed, "Character and Compaction Response of Spray-Dried Agglomerates," *Ceramic Bulletin,* **57** [9] 798-801 (1978).

[8]C.W. Nies and G.L. Messing, "Effect of Glass-Transition Temperature of Polyethylene Glycol-Plasticized Polyvinyl Alcohol on Granule Compaction," *Journal of the American Ceramic Society,* **67** [4] 301-304 (1984).

[9]A.B. van Groenou, "Compaction of Ceramic Powders," *Powder Technology,* **28** 221-228 (1981).

[10]W.J. Walker Jr. and J.S. Reed, "Influence of Slurry Parameters on the Characteristics of Spray-dried Granules," *Journal of the American Ceramic Society,* **82** [7] 1711-1719 (1999).

[11]W.A. Ducker, T.J. Senden and R.M. Pashley, "Direct Force Measurements of Colloidal Forces using an Atomic Force Microscope," *Nature,* **353** 239-241 (1991).

[12]J.P. Cleveland, S. Manne, D. Bocek and P.K. Hansma, "A Nondestructive Method for Determining the Spring Constant of Cantilevers for Scanning Force Microscopy," *Review of Scientific Instruments,* **64** 403-405 (1993).

[13]A. Meurk, J. Yanez and L. Bergström, "Silicon Nitride Granule Friction Measurements with an Atomic Force Microscope: Effect of Humidity and Binder Concentration," *Powder Technology,* **119** 241-249 (2001).

[14]K.M. Andersson and L. Bergström, "Oxidation and Dissolution of Tungsten Carbide Powder in Water," *International Journal of Refractory Metals and Hard Materials,* **18** 121-129 (2000).

[15]J. Aveston, "Hydrolysis of Tungsten (VI). Ultracentrifugation, Acidity Measurements, and Raman Spectra of Polytungstates," *Inorganic Chemistry,* **3** 981-986 (1964).

[16]C. Contescu, J. Jagiello and J.A. Schwartz, "Chemistry of Surface Tungsten Species on WO_3/Al_2O_3 Composite Oxides under Aqueous Conditions," *Journal of Physical Chemistry,* **97** 10152-10157 (1993).

[17]W.P. Griffith and P.J.B. Lesniak, "Raman Studies on Species in Aqueous Solutions. Part III. Vanadates, Molybdates, and Tungstates," *Journal of the Chemical Society. A,* 1066-1071 (1969).

[18]D.L. Kepert, "Isopolytungstates," *Progress in Inorganic Chemistry,* 4 199-274 (1962).

[19]L.G. Sillén, "Stability Constants of Metal-Ion Complexes Section 1. Inorganic Ligands", The Chemical Society, London, 1982.

[20]K.M. Andersson and L. Bergström, "DLVO Interactions of Tungsten Oxide and Cobalt Oxide Surfaces Measured with the Colloidal Probe Technique," *Journal of Colloid and Interface Science,* **246** 309-315 (2002).

[21]K.M. Andersson and L. Bergström, "Effect of the Cobalt Ion and Polyethyleneimine Adsorption on the Surface Forces Between Tungsten Oxide and Cobalt Oxide in Aqueous Media," *Journal of the American Ceramic Society,* **85** [10] 2404-2408 (2002).

[22]I.H. Park and E.-J. Choi, "Characterization of Branched Polyethyleneimine by Laser Light Scattering," *Polymer,* **37** 313-319 (1996).

[23]A. Pfau, W. Schrepp and D. Horn, "Detection of Single Molecule Adsorption Structure of Poly(ethyleneimine) Macromolecules by AFM," *Langmuir,* **15** 3219-3225 (1999).

[24]T. Radeva and I. Petkanchin, "Electric Properties and Conformation of Polyethyleneimine at the Hematite-Aqueous Solution Interface," *Journal of Colloid and Interface Science,* **196** 87-91 (1997).

[25]S.J. Lukasiewicz, "Spray-Drying Ceramic Powders," *Journal of the American Ceramic Society,* **72** [4] 617-624 (1989).

[26]J.S. Reed, "Introduction to the Principles of Ceramic Processing", 1st ed. Wiley, New York, 1988.

[27]S. Baklouti, T. Chartier and J.F. Baumard, "Binder Distribution in Spray-Dried Alumina Agglomerates," *Journal of the European Ceramic Society,* **18** 2117-2121 (1998).

[28]Y. Zhang, X. Tang, N. Uchida and K. Uematsu, "Binder Surface Segregation During Spray Drying of Ceramic Slurry," *Journal of Materials Research,* **13** [7] 1881-1887 (1998).

COLLOIDAL PROCESSING AND LIQUID PHASE SINTERING OF SiC – Al$_2$O$_3$ – Y^{3+} IONS SYSTEM

Nobuhiro Hidaka and Yoshihiro Hirata
Department of Advanced Nanostructured Materials Science and Technology,
Kagoshima University, 1-21-40 Korimoto, Kagoshima 890-0065, Japan

ABSTRACT

An SiC powder of median size 0.8 μm was mixed with an Al$_2$O$_3$ powder of median size 0.2 μm in a 0.3 M-Y(NO$_3$)$_3$ solution at pH 5 to distribute homogeneously the sintering additives (Al$_2$O$_3$+Y^{3+} ion) around the SiC particles. The network structure of SiC particles was formed by the heterocoagulation through the adsorbed Al$_2$O$_3$ and Y^{3+} ion. The aqueous suspension of the SiC-Al$_2$O$_3$ (1.17 – 3.87 vol%)-Y^{3+} ion (0.94 vol% Y$_2$O$_3$) system was consolidated to form green compacts of 52-55 % of theoretical density. These compacts were hot–pressed at 1850° -1950°C to relative density of 95 – 99 %, showing 565– 666 MPa of average flexural strengths and 5.0 – 6.5 MPa•m$^{1/2}$ of fracture toughness.

INTRODUCTION

Silicon carbide (SiC) is potentially useful as a high temperature structural material because of its high strength, high hardness, high creep resistance, and high oxidation resistance. Recently the chemical methods for the addition of sintering additives such as Al$_2$O$_3$ plus Y$_2$O$_3$ to SiC powder have been studied to control the liquid phase sintering and the resultant microstructures of SiC ceramics[1–14]. As compared with the SiC-Al$_2$O$_3$ system, the SiC-Al$_2$O$_3$-

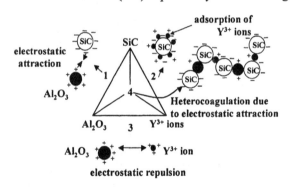

Fig. 1 Possible interactions in aqueous suspension of the SiC-Al$_2$O$_3$-Y^{3+} ion system.

Y_2O_3 system can be densified at a lower sintering temperature, which depends on the Al_2O_3/Y_2O_3 ratio. In our previous papers[3,4,6], the important interactions in the $SiC-Al_2O_3-Y^{3+}$ ion system were investigated to densify with a small amount of sintering additives. Figure 1 shows possible interactions in the aqueous suspension of the three component system. Both the SiC and Al_2O_3 were treated as a form of powder and Y_2O_3 was added as Y^{3+} ion. [1] The interactions between SiC and Al_2O_3 depend on the isoelectric points of the powders and pH of the suspension. In the pH range between the two isoelectric points (pH 2.8 for SiC and pH 7.7 for Al_2O_3), SiC and Al_2O_3 particles were coagulated by the electrostatic attraction. [2] Y^{3+} ions of 0.854 mg/m^2 were adsorbed on the negatively charged SiC surfaces in a $0.3M-Y(NO_3)_3$ solution at pH 5.0. When the pH of the $Y(NO_3)_3$ solution was increased above 6.5, a cloudy precipitate of $Y(OH)_3$ was formed. That is, the adsorption of Y^{3+} ions on SiC surfaces was limited in the pH range below 6.0. This result agreed with the previously reported solubility limit of $Y(OH)_3$ as a function of pH[16] . [3] Electrostatic repulsion between the positively charged Al_2O_3 surfaces and Y^{3+} ions, suppressed the adsorption of Y^{3+} ions. [4] The addition of Al_2O_3 (1.17 % against the SiC volume) and Y^{3+} ion (0.94 % Y_2O_3 against the SiC volume) to the SiC suspension at pH 5.0 formed the network structure of SiC particles by the heterocoagulation through the adsorbed Al_2O_3 and Y^{3+} ion. In this paper, the rheological properties of colloidal suspensions and the sinterability of consolidated powder cake were studied on the $SiC-Al_2O_3-Y^{3+}$ ions system.

EXPERIMENTAL PROCEDURE

The following α-SiC powder, supplied by Yakushima Electric Industry Co., Ltd., Kagoshima, Japan, was used in this experiment : chemical composition, SiC 98.90 mass%, SiO_2 0.66 mass%, C 0.37 mass%, Al 0.004 mass%, Fe 0.013 mass%, median size 0.8 μm, specific surface area 13.4 m^2/g (diameter of equivalent spherical particle 0.14 μm). As sintering additive, α-Al_2O_3 powder of median size 0.2 μm (purity > 99.99 mass% Al_2O_3, specific surface area 10.5 m^2/g, Sumitomo Chemical Industry Co., Ltd., Tokyo, Japan) was added to the SiC powder. The zeta potentials of α-SiC and α-Al_2O_3 powders were measured as a function of pH at a constant ionic strength of 0.01M of NH_4NO_3 (Rank Mark II, Rank Brothers Ltd., Cambridge, UK). The as-received α-SiC powder was dispersed at 30 vol% solids in a $0.3M-Y(NO_3)_3$ aqueous solutions at pH 5.0, and then mixed with the Al_2O_3 powder. The volume ratio of the SiC / Al_2O_3 / Y_2O_3 components was adjusted to 1 / 0.0117 / 0.0094 (1 / 0.0145 / 0.0145 mass ratio). In some experiments, the amount of Al_2O_3 was increased to SiC / Al_2O_3 / Y_2O_3 = 1 / 0.0387 / 0.0094 (1 / 0.0481 / 0.0145 mass ratio). $0.01M-HNO_3$ and $0.01M-NH_4OH$ solutions were used for pH adjustment. The suspensions of the $SiC-Al_2O_3-Y^{3+}$ ion system were stirred by a magnetic stirrer for 12 h and consolidated by filtration through a gypsum mold. The rheological behavior of suspensions was measured by a cone and plate type viscometer (EMD, EHD type, Tokyo Keiki Co., Tokyo, Japan). The sizes of agglomerated SiC particles in the suspensions were also measured by particle size analyzer (CAPA-700, Horiba Ltd., Japan). The consolidated green compacts were pressureless-sintered or hot-pressed under a pressure of 39 MPa at 1850° and 1950°C for 2 h in an Ar atmosphere. The densities of sintered samples were measured by the Archimedes method using kerosene. The surface of sintered SiC ceramics was polished with 1 μm diamond paste and etched with the mixture of NaCl / NaOH = 85/15 (molar ratio) at 800°C for 20 min in air to observe the microstructures by scanning electron microscope (SM300, Topcon Technologies, Inc., Tokyo, Japan). The hot–pressed and pressureless-sintered samples were cut into the specimens with sizes of 3 mm height, 4 mm width and 38 mm length. The specimens were polished with SiC paper of Nos. 600 and 2000 and diamond paste of 6 and 1μm. The flexural strength of SiC-Al$_2$O$_3$-Y^{3+} ion system was measured at room temperature by the four-point flexural method over spans of 30 mm (lower span) and 10 mm (upper span) at a crosshead speed of 0.5 mm/min. The strain gauge was attached on the tensile plane of specimen to measure the Young's modulus. The fracture toughness was evaluated by single–edge V–notch beam (SEVNB) method. A thin diamond blade 1mm thick, where the tip of V–notch had a curvature radius of 20 μm, was used to introduce V-notch of a / W = 0.1 – 0.6 (a : notch length, W : width of the beam). The strength of the notched specimen was measured by three-point loading over a span 30 mm at a crosshead speed of 0.5 mm/min. Equation (1) provides the fracture toughness and equation (2) indicates the shape factor (Y) of crack at S / W =7.5,

$$K_{IC} = \frac{3PS}{2BW^2} Y\sqrt{a} \qquad (1)$$

$$Y = 1.964 - 2.837\lambda + 13.711\lambda^2 - 23.250\lambda^3 + 24.129\lambda^4, \qquad \lambda = \frac{a}{W} \qquad (2)$$

where S, P, and B are the span width, applied load and the thickness of beam, respectively.

RESULTS AND DISCUSSION
Properties of the SiC-Al$_2$O$_3$-Y^{3+} ion suspensions
Figure 2 shows zeta potentials of the SiC and Al$_2$O$_3$ particles as a function of pH. The isoelectric points of SiC and Al$_2$O$_3$ were pH 2.8 and 7.7, respectively. The surface of SiC particle was coated by thin SiO$_2$ film [2]. At a pH below the isoelectric point, the number of positively charged SiOH$_2^+$ sites becomes greater than that of negatively charged SiO$^-$ sites. The opposite case occurs at a pH above the isoelectric point. Similary, the number of AlOH$_2^+$ sites becomes greater than that of AlO$^-$ sites at a pH below the isolelectric point. In this experiment, the suspension of the SiC-Al$_2$O$_3$-Y^{3+} ion system was prepared at pH 5.0. The zeta potentials in Fig. 2 support the electrostatic adsorption of Y^{3+} ion on the negatively charged SiC surface at pH 5.0. On the other hand, the concentrations of Y^{3+} ion between the original 0.3 M-Y(NO$_3$)$_3$ solution and the filtrate separated from the 10 vol% Al$_2$O$_3$ suspension with 0.3 M-Y(NO$_3$)$_3$ at pH 5, were very close within the experimental error less than 0.07 %, indicating no adsorption of Y^{3+} ion on the positively charged alumina surface [6].

Figure 3 shows the apparent viscosity of the suspension of 30 vol% SiC with 1.17 % Al$_2$O$_3$, 0.94 % Y$_2$O$_3$ (as Y^{3+} ion) and 1.17 % Al$_2$O$_3$-0.94 % Y$_2$O$_3$ (as Y^{3+} ion) against SiC volume at pH 5.0. The apparent viscosity of the non-Newtonian flow increased with the sintering additives in the following order : no additive < Al$_2$O$_3$ < Y^{3+} ion < Al$_2$O$_3$+Y^{3+} ion. This result suggests that the network of dispersed SiC particles was formed through the heterocoagulation with positively charged Al$_2$O$_3$ particles or Y^{3+} ions.

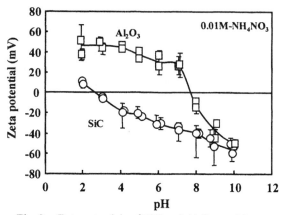

Fig. 2 Zeta potentials of SiC and Al$_2$O$_3$ particles as a function of pH

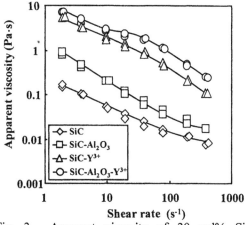

Fig. 3 Apparent viscosity of 30 vol% SiC suspension with 1.17 % Al$_2$O$_3$ and 0.94 % Y$_2$O$_3$ (as Y^{3+} ion) against SiC.

The measured viscosity was closely correlated to the particle size in the suspension. Figure 4 shows the relationship between the apparent viscosity of 30 vol% SiC suspensions at a shear rate of 38.6 s^{-1} and the median sizes measured in the dilute suspensions of 0.0125 mass% solid. The addition of Al$_2$O$_3$, Y^{3+} ion and Al$_2$O$_3$ + Y^{3+} ion caused the agglomeration of dispersed SiC particles (increase of the median size), resulting in the increased apparent viscosity. From the above results, we obtained the following conclusion. The negatively charged monolithic SiC particles were well dispersed due to the strong electrostatic repulsive energy, providing the small median size of particle clusters and the low apparent viscosity of the suspension. The addition of positively charged alumina particles created the hetero-coagulation to the negative-ly charged SiC particles, leading to the increase of the median size of particle clusters and the apparent viscosity. More increase of the median size of particle clusters and the apparent viscosity of the SiC suspension was measured in the addition of Y^{3+} ion, explaining the stronger agglomeration of SiC particles though the adsorbed Y^{3+} ion. The addition of Al$_2$O$_3$ plus Y^{3+} ion to the SiC suspension gave the strongest bridging effects on the agglomeration of SiC particles.

Sintering of the SiC-Al$_2$O$_3$-Y^{3+} ion system

Table I shows the density of the SiC powder compacts consolidated by filtration of the SiC-Al$_2$O$_3$-Y^{3+} ion suspension and calcined at 1000°C in an Ar atmosphere. The

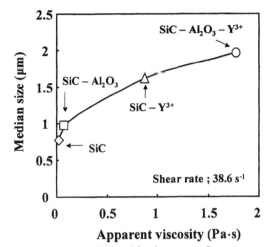

Fig. 4 Relationship between the apparent viscosity of 30 vol% SiC suspension at a shear rate 38.6 s^{-1} and the median size of SiC with Al$_2$O$_3$ and Y^{3+} ion in dilute suspension.

monolithic SiC suspension gave the highest-density compact of 55 % of theoretical density. The decrease of green density with the addition of Al$_2$O$_3$ or Y^{3+} ion was 1.5-3 %. This result suggests that the SiC particles were agglomerated to form a flexible network structure and were compressed by the suction pressure of suspension liquid through gypsum mold. The subsequent isostatic pressing of the consolidated SiC compacts (No. 6 in Table I) was effective to increase the bulk density to 61 % of theoretical density. Figure 5 shows the relation between the density of the SiC compacts after the pressure-less-sintering at 1850° -1950 °C and the amount of the liquid phase, calculated using the phase diagrams of SiO$_2$-Al$_2$O$_3$[17], SiO$_2$-Y$_2$O$_3$[18] and SiO$_2$-Al$_2$O$_3$-Y$_2$O$_3$[19] SiO$_2$-Al$_2$O$_3$-Y$_2$O$_3$[19] systems. The amount of SiO$_2$ component was based on the chemical composition of SiO$_2$ included in the as-received SiC powder. When the SiC dissolves inthe oxide liquid, the amount of liquid phase becomes higher than that of liquid calculated in Fig. 5. However, the exact solubility of SiC in the oxide liquid has not been

Table I. Density of the SiC compacts with 1.17 vol% Al$_2$O$_3$ and 0.94 vol% Y$_2$O$_3$ (as Y^{3+} ions), calcined at 1000 °C

Sample	Sintering additive (vol %)	Bulk density (g / cm³)	Relative density (%)
No.1	None	1.80 (±0.01)	55.3 (±0.6)
No.2	Y^{3+} (0.94% Y$_2$O$_3$)	1.73 (±0.02)	53.8 (±0.5)
No.3	Al$_2$O$_3$ (1.17 %)	1.69 (±0.01)	52.3 (±0.3)
No.4	Al$_2$O$_3$ (1.17 %) + Y^{3+} (0.94% Y$_2$O$_3$)	1.73 (± 0.01)	53.3 (±0.1)
No.5	Al$_2$O$_3$ (3.87 %) + Y^{3+} (0.94% Y$_2$O$_3$)	1.70 (± 0.002)	52.4 (±0.1)
No.6*	Al$_2$O$_3$ (3.87 %) + Y^{3+} (0.94% Y$_2$O$_3$)	1.99 (± 0.01)	61.2 (±0.1)

*) The powder compact consolidated by filtration was subsequently compressed by isostatic pressing under a pressure of 196 MPa.

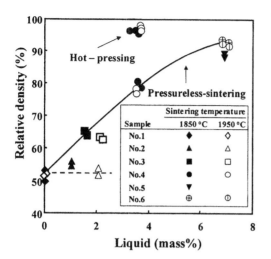

Fig. 5 Density of the SiC compacts after pressureless-sintering or hot-pressing at 1850°-1950°C and the calculated amount of the liquid phase in the SiC-Al_2O_3-Y_2O_3 system.

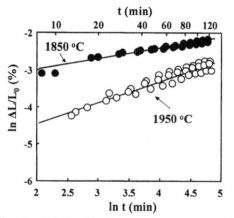

Fig. 6 Relationship between hot-pressing time and shrinkage of SiC at 1850°-1950°C on log-log axes.

reported. No densification of monolithic SiC compact (No. 1) occurred in the pressureless-sintering. Only the addition of Y^{3+} ion (No. 2) has no effect on the densification of SiC. On the other hand, Al_2O_3 component (No.3) enhanced the sinterability of SiC compact through the formation of the liquid of the SiO_2-Al_2O_3 system. The addition of the Al_2O_3-Y_2O_3 (as Y^{3+} ion) components (Nos. 4-6) has a great effect on the densification of SiC and the density of SiC reached 80 % of theoretical density at 3.6 mass% of liquid. When the amount of liquid phase was increased to 7.0 mass%, the density of SiC compacts was enhanced to 93 % of theoretical density. A similar density of SiC sintered at 1850° and 1950°C indicates a small effect of the viscosity of liquid at those high temperatures. Figure 5 shows also the relative density of the SiC compacts with 1.17 vol% (1.45 mass%) Al_2O_3-0.94 vol% (1.45 mass%) Y_2O_3 after the hot-pressing under a pressure of 39 MPa. The shrinkage of SiC compacts started at 1420°C, which was close to the liquid formation temperature (≈1400 °C) at a given composition of the SiO_2(0.66 mass%) - Al_2O_3(1.45 mass%)-Y_2O_3(1.45 mass%) sys-

tem. After the hot-pressing at 1850°-1950°C, the relative density of SiC compact reached 96.6 ± 2.3 % and 96.5 ± 2.1 %, respectively.

Figure 6 shows the relationship between the shrinkage (Δh / ho) of compacts based on the sample height (ho) for 0 h at 1850°-1950°C and hot-pressing time (t) on log-log axes. A good linear relation was observed at each temperature. The slope (k) for the ratio

Fig. 7　Microstructures of SiC compacts with Al_2O_3 and Y^{3+} ion, pressureless-sintered (a-d) and hot-pressed (e,f) at 1850°-1950°C.

of log (Δh / ho) / log t was 0.329 ± 0.047 and 0.691 ± 0.008 at the hot–pressing temperature of 1850° and 1950°C, respectively. The following relationships between the mass transport mechanism and k value are proposed in the early stage sintering with liquid phase : k = 0.33 for liquid phase sintering dominated by diffusion in liquid [20], k = 0.50 for liquid phase sintering dominated by dissolution-reprecipitation mechanism [20], and k = 1.0 for viscous deformation mechanism [21]. The measured result suggests that (1) densification of the SiC compact with Al_2O_3 and Y^{3+} ion proceeded by liquid phase sintering dominated by diffusion of SiC in the liquid at 1850 °C and (2) the SiC compact at 1950 °C was densified by the mixed mass transport mechanisms of dissolution -reprecipitation of SiC and viscous deformation with the liquid.

Microstructures of SiC compacts

Figure 7 shows the microstructures of SiC compacts pressureless-sintered and hot-pressed at 1850° -1950°C. The addition of Al_2O_3 component (a) accelerated the densification with grain growth of SiC. The microstructure (b) of the SiC with Al_2O_3 (1.17 vol%) + Y^{3+} ion (0.94 vol% Y_2O_3) shows the intermediate stage of sintering with grain growth. The increase of the amount of sintering additives (3.87 vol% Al_2O_3 - 0.94 vol% Y_2O_3) accelerated the densification of SiC with grain growth (c). A similar microstructure (d) was observed in the SiC sintered at 1950°C, indicating a small influence of increased sintering temperature on the microstructures. The densification with grain growth for SiC with Al_2O_3 -Y^{3+} ions is associated with the enhanced mass transport in the liquid formed at the grain boundaries or related to the increased driving force of sintering due to the decreased interfacial energy at the grain boundaries. The application of pressure during the sintering enhanced the mass transport contributing to the densification. The average grain sizes, measured on 200 grains, were 2.9 and 3.5 μm for Fig.7(e) and (f), hot-pressed at 1850° and 1950°C, respectively.

Table II. Mechanical properties of the SiC-Al$_2$O$_3$-Y^{3+} system

	Pressureless-sintered at 1850°C (Sample No.6*)		Hot-pressed at 1850°C (Sample No.4*)		Hot-pressed at 1950°C (Sample No.4*)	
Flexural strength range (MPa)	190 (min.)	435 (max.)	369 (min.)	793 (max.)	433 (min.)	637 (max.)
Measured maximum flaw size (μm)	475	30	92	34	35	65
Weibull modulus	4.0		5.9		11.4	

∗) See Table I for samples.

Mechanical properties of SiC-Al$_2$O$_3$-Y$_2$O$_3$ system

Table II summarizes the mechanical properties of the SiC-Al$_2$O$_3$-Y^{3+} system, densified at 1850°-1950 °C. The pressureless-sintered and hot-pressed samples showed a brittle fracture behavior with the Young's moduli of 402-520 GPa. The strength at a failure probability of 50 % and the Weibull modulus resulted in 319 MPa and 4.0 for the 10 specimens pressuless-sintered at 1850 °C, 666 MPa and 5.9 for the 14 specimens hot-pressed at 1850°C, and 565 MPa and 11.4 for the 15 specimens hot-pressed at 1950°C, respectively. The mechanical properties of the SiC with Al$_2$O$_3$ and Y$_2$O$_3$ were improved through the densification by hot-pressing. The fracture toughness of the

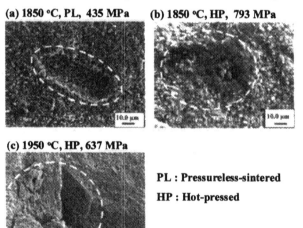

(a) 1850 °C, PL, 435 MPa (b) 1850 °C, HP, 793 MPa

(c) 1950 °C, HP, 637 MPa

PL : Pressureless-sintered

HP : Hot-pressed

Fig. 8   Flaw observed in the fracture surfaces of pressureless-sintered and hot-pressed SiC.

SiC hot-pressed at 1850°- 1950 °C was determined by the SEVNB method to be 5.8 ± 0.9 MPa·m$^{1/2}$. No significant difference in the fracture toughness was measured between the SiC samples hot-pressed at 1850° and 1950°C. That is, the Weibull modulus is strongly related to the flaw size distribution of the SiC. Figure 8 shows the fracture origin of pressureless-sintered and hot-pressed SiC. In the SiC pressureless-sintered at 1850°C, a porous part formed the fracture origin (a). The flaw observed in the SiC hot-pressed at

1850°C (b) was the solidified region of liquid during the cooling process. The flaw formed in the hot-pressed at 1950°C (c) was the porous part.

CONCLUSIONS

In this experiment, the properties of aqueous suspensions of the SiC-Al$_2$O$_3$-Y^{3+} ion system at pH 5.0 was studied to distribute homogeneously the sintering additives around the SiC particles. The powder compacts consolidated through gypsum mold were sintered at 1850°-1950°C in an Ar atmosphere.

(1) Y^{3+} ions of 0.854 mg/m^2 were adsorbed on the negatively charged SiC surface with isoelectric point pH 2.8 in a 0.3M-Y(NO$_3$)$_3$ solution. Electrostatic repulsion between the positively charged Al$_2$O$_3$ surface with isoelectric point pH 7.7 and Y^{3+} ion, suppressed the adsorption of Y^{3+} ion.

(2) The addition of Al$_2$O$_3$ (1.17 % against the SiC volume) and Y^{3+} ion (0.94 % Y$_2$O$_3$ against the SiC volume) to the SiC suspension at pH 5.0 formed the network structure of SiC particles by the heterocoagulation through the adsorbed Al$_2$O$_3$ and Y^{3+} ion. The size of agglomerated particles and apparent viscosity of the suspension increased in the following order : no additive < Al$_2$O$_3$ < Y^{3+} ion < Al$_2$O$_3$ + Y^{3+} ion.

(3) The green density of the consolidated SiC compacts with and without the sintering additives was 52-55 % of theoretical density. The bulk density of SiC increased with an increase in the amount of the liquid phase in the pressureless-sintering at 1850°-1950°C. The densification of the SiC was enhanced with grain growth in the following order of sintering additives : no additive ≈ Y^{3+} ion < Al$_2$O$_3$ < Al$_2$O$_3$ + Y^{3+} ion.

(4) When a pressure of 39 MPa was applied to the SiC compact with Al$_2$O$_3$ and Y^{3+} ion, the relative density was greatly increased to 95-99 % at 1850°-1950°C. The SiC compacts were densified with grain growth by the mixed mass transport mechanisms of dissolution–reprecipitation of SiC and viscous deformation by the liquid phase formed at 1950°C.

(5) The strength at a failure probability of 50 % and the Weibull modulus of SiC resulted in 319 MPa and 4.0 for the pressureless-sintering at 1850°C, 666 MPa and 5.9 for the hot-pressing at 1850 °C and 565 MPa and 11.4 for the hot-pressing at 1950 °C, respectively.

REFERENCES

[1] E. Liden, E. Carlstrom, L. Eklund, B. Nyberg and R. Carlsson, "Homogeneous Distribution of Sintering Additives in Liquid-phase Sintered Silicon Carbide," J. Am. Ceram. Soc, 78, 1761-1768 (1995).

[2] Y. Hirata, K. Miyano, S. Sameshima and Y. Kamino, "Reaction between SiC Surface and Aqueous Solutions Containing Al Ions,"Colloid and Surfaces, A : Physicochem. & Eng. Aspects, 133, 183-189 (1995).

[3] Y. Hirata, S. Tabata and J. Ideue, "Interaction of the Silicon Carbide-Polyacrylic Acid-Yttium ion System," J. Am. Ceram. Soc., 86, 5-11 (2003).

[4] Y. Hirata, K. Hidaka, H. Matsumura, Y. Fukushige and S. Sameshima, "Colloidal Processing and Mechanical Properties of Silicon Carbide with Alumina," J. Mater. Res., 12, 3146-3157 (1997).

[5]S. Sameshima, K. Miyano and Y. Hirata, "Sinterability of SiC Powder Coated Uniformly with Al Ions," J. Mater. Res, 13, 816-820 (1998).

[6]N. Hidaka, Y. Hirata and S. Sameshima, "Colloidal Processing of the SiC-Al$_2$O$_3$-Y^{3+} Ions System and Sintering Behavior of the Consolidated Powder Compacts," J. Ceram. Proc. Res., 3, 271-277 (2002).

[7]M. A. Mulla and V. D. Krstic, "Low-temperature Pressureless Sintering of β-Silicon Carbide with Alumina Oxide and Yttrium Oxide Additions," Am. Ceram. Soc. Bull., 70, 439-443 (1991).

[8]D. Sciti, and A. Bellosi, "Effects of Additives on Densification, Microstructure and Properties of Liquid-phase Sintered Silicon Carbide," J. Mater. Sci., 35, 3849-3855, (2000).

[9]L. M. Wang and W. C. Wei, "Colloidal Processing and Liquid-phase Sintering of SiC," J. Ceram. Soc. Jpn., 103, 434-443 (1995).

[10]J. H. She and K. Ueno, "Densification Behavior and Mechanical Properties of Pressureless-sintered Silicon Carbide Ceramics with Alumina and Yttria Additions," Mater. Chem. Phys., 59, 139-142 (1999).

[11]V. D. Krstic, "Optimization of Mechanical Properties in SiC by Control of the Microstructure," Mater. Res. Soc. Bull., XX, 46-49 (1995).

[12]J. H. She and K. Ueno, "Effect of Additive Content on Liquid-phase Sintering on Silicon Carbide Ceramics," Mater. Res. Bull., 34, 1629-1636(1999).

[13]Y. W. Kim, J. Y. Kim, S. H. Rhee and D. Y. Kim, "Effect of Initial Particle Size on Microstructure of Liquid-phase Sintered α-Silicon Carbide," J. Eur. Ceram. Soc., 20, 945-949 (2000).

[14]V. V. Pujar, R. P. Jensen and N. P. Padture, "Densification of Liquid-phase- sintered Silicon Carbide," J. Mater. Sci. Lett., 19, 1011-1014 (2000).

[15]Y. Hirata and W. H. Shih, "Colloidal Processing of Two-component Powder System" ; pp. 637-644 in Advances in Science and Technology 14, Proceedings of 9 th Cimtec-World Ceramics Congress, Ceramics : Getting into the 2000′s-Part B, Edited by P. Vincenzini, Techna Srl.,Faenza, Italy, 1999.

[16]Jr. C. F. Baes and R. E. Mesuner, p.134 in The Hydrolysis of Cations, Robert E. Krieger Pub., Malabar, Florida, 1986.

[17]I. A. Aksay and J. A. Pask, Stable and Metastable Equilibria in the System SiO$_2$-Al$_2$O$_3$, " J. Am. Ceram. Soc., 58, 507-512 (1975).

[18]E. M. Levin, C. R. Robbins and H. F. McMurdie, Fig. 2388 in Phase Diagram for Ceramists, The American Ceramic Society, Columbus, Ohio, 1969.

[19]E. M. Levin, C. R. Robbins, and H. F. McMurdie, Fig. 2586 in Phase Diagram for Ceramists, The American Ceramic Society, Columbus, Ohio, 1969.

[20]W. D. Kingery, "Densification During in the Presence of a Liquid Phase. I. Theory," J. Appl. Phys., 30, 301-306 (1959).

[21]W. D. Kingery and M. Berg, "Study of the Initial Stage of Sintering Solids by Viscous Flow, Evaporation-condensation, and Self-diffusion," J. Appl. Phys., 26, 1205-1212 (1956).

COLLOIDAL PROCESSING OF SiC WITH 700 MPa OF FLEXURAL STRENGTH

Shuhei Tabata and Yoshihiro Hirata
Department of Advanced Nanostructured Materials Science and Technology, Kagoshima University
1 - 21 - 40 Korimoto, Kagoshima 890 – 0065, Japan

ABSTRACT

An SiC powder of median size 0.8 μm was mixed with polyacrylic acid (PAA, dispersant) in a 0.3 M-Y(NO$_3$)$_3$ solution at pH 2-6 to adsorb uniformly the sintering additive (Y^{3+} ions) on the SiC surface. The addition of PAA to the SiC suspension with Y^{3+} ions increased the amount of Y^{3+} ions fixed to SiC particles by the two types of fixation mechanisms of Y^{3+} ions : the accumulation effect by neutral PAA and the electrostatic attraction effect by negatively charged PAA. The aqueous 30 vol% SiC suspension with 0.52 % PAA and 1.53 % Y$_2$O$_3$ (as Y^{3+} ions) against the mass of SiC, was consolidated by filtration through a gypsum mold to form green compacts of 51 % of theoretical density. The green compacts were densified with grain growth to 94-98 % relative density by the dissolution-reprecipitation mechanism of SiC in the SiO$_2$-Y$_2$O$_3$ liquid during the hot-pressing at 1950°C. The dense SiC ceramics showed the excellent mechanical properties : 550-910 MPa of four-point flexural strength, 6.7 of Weibull modulus, 5.0-7.8 MPa·m$^{1/2}$ of fracture toughness and 20 GPa of Vickers hardness.

INTRODUCTION

Silicon carbide (SiC) is potentially useful as a high temperature structural material because of its high strength, high hardness, high creep resistance and high oxidation resistance. For the fabrication of dense SiC, sintering additives of metal compounds, or boron and carbon, are needed owing to the high covalency of SiC. Recently the chemical methods for the addition of sintering additives such as Al$_2$O$_3$ plus Y$_2$O$_3$ to SiC powder have been studied to control the liquid phase sintering and the resultant microstructures of SiC ceramics[1-4]. The sinterability of the SiC-Al$_2$O$_3$-Y$_2$O$_3$ system depends on the amount and ratio of

Al_2O_3-Y_2O_3[5-8]. The densification of SiC progresses through the dissolution-reprecipitation mechanism of SiC in the liquid phase of the SiO_2-Al_2O_3-Y_2O_3 system. The SiO_2 component is naturally formed on the surface of as-received SiC powder[9]. The chemical methods of SiC with the sintering additives are expected to provide the following advantages : (1) a homogeneous distribution of the additives around SiC particles, (2) the increased densification rate by the well-distributed liquid and (3) the decrease of the amount of additives[10]. In our previous paper[11], we investigated the interactions of the silicon carbide-yttrium ions system in aqueous suspension to disperse homogeneously the sintering additive around SiC particles. The derived conclusions are as follows : (1) The amount of Y^{3+} ions adsorbed on SiC particles of isoelectric point pH 2.5, which was represented by the Langmuir adsorption isotherm[12, 13], increased with an increase of pH because of the electrostatic attraction between negatively charged SiC surface and Y^{3+} ions. (2) The adsorption of Y^{3+} ions onto SiC surface enhanced the coagulation of SiC particles through the electrostatic attraction between negatively charged SiC surfaces and Y^{3+} ions fixed. As a result, the apparent viscosity of SiC suspension was increased. In this paper, we report on the interactions of SiC-Y^{3+} ions-polyacrylic acid (PAA) system to prepare a well-dispersed SiC suspension with the sintering additive. Furthermore, the sinterability, microstructure and mechanical properties of the SiC-Y_2O_3 compacts densified by hot-pressing, were studied. The liquid formation temperature of the SiO_2 (surface layer of SiC)-Y_2O_3 systems is reported to be 1775°C[15]. This temperature is 300°C higher than the liquid formation temperature of the SiO_2-Al_2O_3-Y_2O_3 system. Once the SiC-Y_2O_3 system is highly densified, the thermal resistance of SiC is greatly enhanced.

EXPERIMENTAL PROCEDURE

An α-SiC powder with the following characteristics, supplied by Yakushima Electric Industry Co., Ltd., Kagoshima, Japan, was used in this experiment : chemical composition, SiC 98.90 mass%, SiO_2 0.66 mass%, C 0.37 mass%, Al 0.004 mass%, Fe 0.013 mass%, median size 0.8 μm, specific surface area 13.4 m^2/g. As-received α-SiC powder was dispersed at 30 vol% solid in 0.3 M of Y $(NO_3)_3$ aqueous solutions at pH 2-5 and then polyacrylic acid (PAA, average molecular weight 10000, Daiichi Kogyo Co., Kyoto, Japan) of 0.39-0.60 mg/m^2-SiC surface (saturated amount) was added to be adsorbed on the SiC surface. The amount of Y^{3+} ions adsorbed on the SiC particles was determined by chelatemetric titration of Y^{3+} ions in the filtrate separated from SiC suspension. The amount of PAA adsorbed on SiC particles was determined by measuring the weight loss upon heating of the PAA-adsorbed SiC powder to 800°C in an Ar atmosphere (Thermo Plus TG-8210, Rigaku Co., Japan). 0.1 M - HNO_3 and 0.1 M - NH_4OH solutions were used for pH adjustment. When the pH of the $Y(NO_3)_3$ solution was increased above 6.5, a cloudy precipitate of $Y(OH)_3$ was formed.

This result agreed with the previously reported solubility limit of Y(OH)$_3$ as a function of pH[17]. The volume ratio of the SiC / Y$_2$O$_3$ (as Y^{3+} ions) was adjusted to 1 / 0.01 (1 / 0.015 mass ratio). The zeta potential of Y^{3+} ions - adsorbed SiC particles and PAA - adsorbed SiC particles were measured at a constant ionic strength of 0.01 M NH$_4$NO$_3$ (Rank Mark II, Rank Brothers Ltd., Cambridge, UK). The rheological behavior of the suspensions was measured by cone and plate type viscometer (EHD type, Tokimec Inc., Tokyo, Japan). The suspensions were consolidated by filtration through a gypsum mold. Some consolidated powder compacts were isostatically compressed under a pressure of 196 MPa. Green compacts were calcined at 1000°C for 1 h in an Ar atmosphere to measure the density.

The consolidated green compacts were hot–pressed under a pressure of 39 MPa at 1750° - 1950°C for 2 h in an Ar flow. The heating and cooling rates were 10°C/min. The densities of sintered samples were measured by the Archimedes method using kerosene. The surface of sintered SiC ceramics was polished with 1 μm diamond paste and etched with the mixture of NaCl / NaOH = 85/15 (molar ratio) at 550°C for 5 min in air to observe the microstructures by scanning electron microscope (SM300, Topcon Technologies, Inc., Tokyo, Japan). The hot-pressed SiC was cut into the specimens with sizes of 3 mm height, 4 mm width and 38 mm length. The specimens were polished with SiC paper of Nos. 600 and 2000 and diamond paste of 6 and 1 μm. The Vickers hardness of the densified SiC was measured at 9.8 N. The flexural strength of SiC-Y^{3+} ion system was measured at room temperature by the four-point flexural method over spans of 30 mm (lower span) and 10 mm (upper span) at a crosshead speed of 0.5 mm/min. The strain gauge was attached on the tensile plane of specimen to measure the Young's modulus. The fracture toughness was evaluated by single-edge V-notch beam (SEVNB) method. A thin diamond blade 1mm thick, where the tip of V-notch had a curvature radius of 20 μm, was used to introduce V-notch of a / W = 0.1 - 0.6 (a : notch length, W : width of the beam). The strength of the notched specimen was measured by three-point loading over a span 30 mm at a crosshead speed of 0.5 mm/min. Equation (1) provides the fracture toughness and equation (2) indicates the shape factor (Y) of crack at S / W =7.5. S, P, and B in Eqs. (1) and (2) are the span width, applied load and the thickness of beam, respectively.

$$K_{IC} = \frac{3PS}{2BW^2} Y\sqrt{a} \tag{1}$$

$$Y = 1.964 - 2.837\lambda + 13.711\lambda^2 - 23.250\lambda^3 + 24.129\lambda^4 \quad (\lambda = \frac{a}{W}) \tag{2}$$

RESULTS AND DISCUSSION
Adsorption of Y^{3+} ions and PAA onto SiC surface

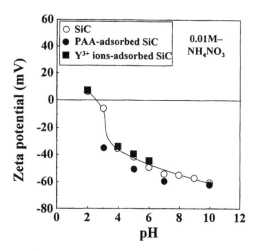

Fig. 1 Zeta potential of SiC, Y^{3+}ions-adsorbed SiC and PAA-adsorbed SiC.

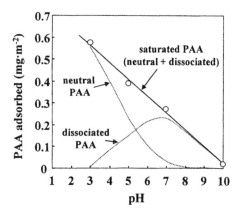

Fig. 2 Saturated amount of PAA adsorbed on SiC surfaces and fraction of dissociated and neutral polymer as a function of pH.

Figure 1 shows the zeta potential of SiC suspensions as a function of pH. The surfaces of SiC particles were coated by thin SiO_2 films[9]. At a pH below the isoelectric point (pH 2.5), the number of positively charged $SiOH_2^+$ sites becomes greater than that of negatively charged SiO^- sites. The opposite case occurs at a higher pH above the isoelectric point. The zeta potential of SiC was slightly changed toward the positive value with adsorption of Y^{3+} ions. Adsorption of PAA onto SiC surface shifted the zeta potential of SiC particles to the negative value. The saturated amount of PAA adsorbed is shown in Fig. 2, indicating a linear decrease in the saturated value at a higher pH. This result is explained by the surface charge of SiC and the dissociation of PAA with pH. The dissociation of PAA starts at about pH 3, and reaches 100 % above pH 9[16]. Based on our previous study[16], the fractions of neutral PAA and negatively charged PAA are also shown in Fig. 2. The decrease in the saturated amount of polymer with increasing pH is due to the electrostatic repulsion between the negatively charged polymer and the negatively charged SiC surface. The increased zeta potential of SiC to negative values with addition of PAA (Fig. 1) indicates the adsorption of negatively charged PAA on the positively charged $SiOH_2^+$ sites. In the neutral PAA, the localization of electrons in carboxyl group produces $O^{\delta-}$ and $C^{\delta+}$ atoms. This phenomenon enhances the interaction between $SiOH_2^+$ sites of SiC surface and $O^{\delta-}$ atoms in

Colloidal Ceramic Processing

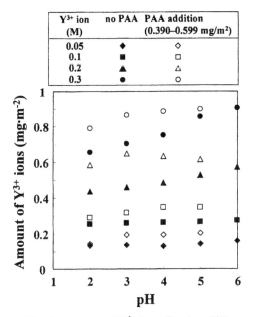

Y^{3+} ion (M)	no PAA	PAA addition (0.390–0.599 mg/m²)
0.05	◆	◇
0.1	■	□
0.2	▲	△
0.3	●	○

Fig. 3 Amount of Y^{3+} ions fixed to SiC surface in suspensions with and without PAA.

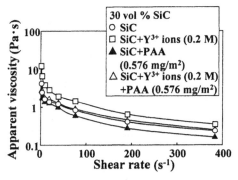

30 vol % SiC
○ SiC
□ SiC+Y^{3+} ions (0.2 M)
▲ SiC+PAA
(0.576 mg/m²)
△ SiC+Y^{3+} ions (0.2 M)
+PAA (0.576 mg/m²)

Fig. 4 Effects of addition of Y^{3+} ions and PAA of the apparent viscosity of SiC suspensions at pH 3.0.

PAA and between SiO⁻ sites and $C^{\delta+}$ atoms in PAA, explaining the increased amount of PAA fixed to SiC particles at a lower pH.

The amount of Y^{3+} ions adsorbed on SiC particles increased with an increase of pH because of the electrostatic attraction between negatively charged SiC surfaces and positively charged Y^{3+} ions (Fig.3). The increase of concentration of $Y(NO_3)_3$ solutions was another factor to enhance the amount of adsorbed. Figure 3 shows the amount of Y^{3+} ions fixed to the SiC particles in the suspensions with the saturated amounts of PAA in the pH range from 2 to 5. The addition of PAA increased the amount of Y^{3+} ions fixed, indicating both the adsorption of Y^{3+} ions to the surfaces of SiC particles and to the fixed PAA. The data in Fig. 3 suggest the following two types of fixation mechanisms of Y^{3+} ions through adsorbed PAA : (1) accumulation effect by neutral PAA. The amount of Y^{3+} ions fixed to PAA adsorbed was influenced by pH and the concentration of Y^{3+} ions. In the SiC suspensions at pH 2 - 3, the neutral PAA is fixed to the lowly charged SiC surfaces near the isoelectric point. The data in Fig. 3 indicate the accumulation of Y^{3+} ions in the neutral polymer layer adsorbed on the SiC surfaces at pH 2-3. This effect decreases with increasing pH, because the fraction of neutral polymer decreases with pH. (2) Electrostatic attraction effect by negatively charged PAA. With an increase of pH of the SiC suspension, the

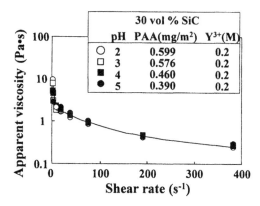

Fig. 5 Influence of pH on the apparent viscosity of SiC suspensions containing PAA and Y^{3+} ions.

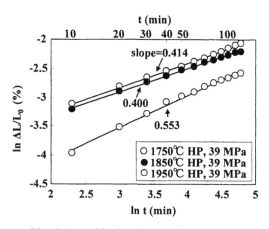

Fig. 6 Logarithmic relationship between shrinkage of SiC compact with Y^{3+} ions and hot-pressing time.

charged polymer is adsorbed at a high pH in the expanded structure due to the electrostatic repulsion between charged carboxylate ions. The electrostatic attraction effect becomes greater at a higher pH, because the dissociation of polymer proceeds with increasing pH.

Rheological properties of SiC suspensions

Figure 4 shows the influence of Y^{3+} ions addition on the rheological properties of SiC suspensions of 30 vol% solid at pH 3.0. The addition of Y^{3+} ions to negatively charged SiC particles increased the viscosity slightly because of the coagulation of SiC particles through the adsorbed Y^{3+} ions. On the other hand, PAA addition decreased the viscosity of the SiC suspension by the steric stabilization effect of neutral PAA. Apparently, Y^{3+} ions addition and PAA addition have an opposite influence on the viscosity of the SiC suspension. As a result, a small change in the viscosity of the SiC suspension was measured for both the addition of Y^{3+} ions and PAA. Figure 5 shows the influence of pH on the apparent viscosity of SiC suspensions with Y^{3+} ions and the saturated amount of PAA adsorbed. The increased pH promotes the dissociation of PAA. The PAA fixed to the SiC surfaces is expected to give the electrosteric stabilization effect to the SiC particles. However, the difference in viscosity of the PAA - containing SiC suspensions with pH was relatively small.

Hot-pressing of the SiC-Y^{3+} ions compact

The SiC powder were consolidated by filtration of the SiC-Y^{3+} ions-PAA

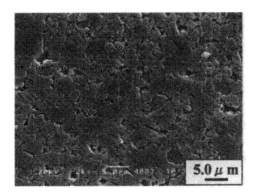

Fig. 7 Microstructure of the SiC with Y^{3+} ions, hot-pressed at 1950°C.

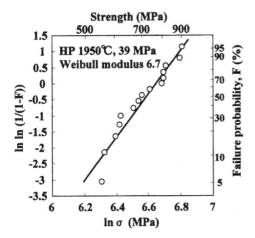

Fig. 8 Weibull plot of flexural strength of SiC hot-pressed with 1 vol % Y_2O_3 (as Y^{3+} ions) at 1950°C.

suspension at pH 5.0 and calcined at 1000°C in Ar atmosphere. The packing density was 49-53 % of theoretical density. Little influence of the amount of Y^{3+} ions adsorbed was measured on the packing density. The subsequent isostatic pressing of the consolidated SiC compacts was effective to increase the bulk density to 60 % of theoretical density. The relative density of the SiC compacts with 1 vol% (1.53 mass%) Y_2O_3 after the hot-pressing under a pressure of 39 MPa at 1750°, 1850° and 1950°C reached 80.0 ± 2.1%, 85.5 ± 0.5 % and 96.2 ±1.8 %, respectively. The relationship between the shrinkage ($\Delta L / Lo$) of compacts based on the sample length (Lo) for 0 h at 1750° - 1950°C and hot - pressing time (t) was plotted on log-log axes in Fig. 6. A good linear relation was observed at each hot- pressing temperature. The slope (k) was 0.414 ± 0.055 at 1750°C, 0.400 at 1850°C and 0.553 ± 0.114 at 1950°C. The following relation-ships between the mass transport mechanism and k value are proposed in the early stage sintering : k= 0.33 for mass transport by grain boundary diffusion and for liquid phase sintering dominated by diffusion in liquid[17], k=0.40 for mass transport by lattice diffusion, k = 0.50 for liquid phase sintering dominated by dissolution-reprecipitation mechanism[17], and k = 1.0 for viscous deformation mechanism[18]. The measured result suggests that (1) the densification of SiC at 1750°-1850°C proceeded through the mass transport by lattice diffusion, which was enhanced by the adsorbed Y^{3+} ions, and (2) the densification at 1950°C was due to the liquid phase sintering of the SiC-SiO₂-Y_2O_3 system dominated by dissolution-reprecipitation of SiC. Figure 7 shows the microstructure of SiC

Table I. Comparison of mechanical properties of dense SiC.

	Ref. 19	Ref. 19	This work
Sintering temperature (°C)	2000	2000	1950
Additive	boron	alumina	yttria
Density (g/cm³)	3.15	3.20	3.14
Young's modulus (GPa)	410	455	492±33
Weibull modulus (Number of specimens)	9.3 (65)	27 (30)	6.7 (15)
Mean failure strength (MPa)	370	560	719

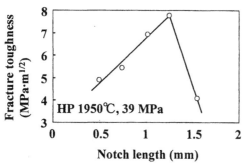

HP 1950℃, 39 MPa

Notch length (mm)

Fig. 9 Relationship between fracture toughness and notch length of hot-pressed SiC in SEVNB method.

hot-pressed with 1 vol% Y_2O_3 (as Y^{3+} ions) at 1950°C for 2 h. The average grain size, measured on 200 grains, was 2.4 μm. Since the median size of starting SiC particles was 0.8 μm, the densification of SiC was accompanied by the grain growth.

Mechanical properties of the hot–pressed SiC-Y_2O_3 system

The SiC samples hot-pressed at 1950°C showed a brittle fracture behavior with the Young's moduli of 459-545 GPa. Figure 8 shows the Weibull plot of four point flexural strength of the SiC with Y_2O_3. The minimum and maximum strengths for 15 specimens were 549 and 908 MPa, respectively. The strength at a failure probability of 50% and the Weibull modulus resulted in 719 MPa and 6.7, respectively. Table I compares the mechanical properties of dense SiC prepared in this experiment with those reported. As compared to SiC with B or Al_2O_3, Y_2O_3 addition increased the Youngs' modulus and fracture strength. The Weibull modulus in this experiment was relatively low. Figure 9 shows the relationship between the fracture toughness and the notch length of SEVNB method for the SiC hot-pressed at 1950°C. The fracture toughness of the SiC depended on the notch length and was in the range from 5.0 to 7.8 MPa·m$^{1/2}$. The notch size dependance of fracture toughness is also reported by Trentini el al. in the SB (sandwiched beam) method[20]. The reason for this result is under investigation. The crack formed by Vickers indenter in the hot-pressed SiC propagated along the grain boundary, causing the crack deflection and bridging by grains. The flaw sizes in the SiC at a failure probability of 10, 50 and 90% in Fig. 8 was estimated by Eqs. (1) and (2) using the fracture toughness of 5 and 7 MPa·m$^{1/2}$ as follows : (i) K_{IC}=5 MPa·m$^{1/2}$, flaw size : 22 μm/10%, 13 μm/50% and 8 μm/90%, (ii) K_{IC}=7 MPa·m$^{1/2}$, flaw size : 43 μm/10%, 26 μm/50% and 16 μm/90%. This calculation suggests that small flaws were formed by the present processing. The fractured surface of hot-pressed SiC was observed to identify the fracture origin but it was difficult to see it. The Vickers hardness at a load of 9.8 N was 20.0±0.07 GPa. This value was comparable to the reported values[7, 8, 21,].

CONCLUSIONS

(1) The amount of Y^{3+} ions adsorbed on SiC particles of isoelectric point pH 2.5, increased with an increase of pH because of the electrostatic attraction between negatively charged SiC surface and Y^{3+} ions.

(2) The amount of PAA adsorbed on SiC particles decreased with increasing pH because of the electrostatic repulsion between negatively charged SiC surface and dissociated PAA.

(3) The addition of PAA to the SiC suspensions with Y^{3+} ions increased the amount of Y^{3+} ions fixed to SiC particles. This result was explained by the two types of fixation mechanisms of Y^{3+} ions through adsorbed PAA : accumulation effect by neutral PAA and electrostatic attraction effect by negatively charged PAA.

(4) The SiC suspension with Y^{3+} ions and PAA was consolidated to 49-53 % of theoretical density by filtration through gypsum mold. When a pressure of 39 MPa was applied to the SiC compact with Y^{3+} ions, the relative density was greatly increased to 94-98 % at 1950°C. The SiC compacts were densified with grain growth by the enhanced mass transport through the liquid phase sintering dominated by dissolution-reprecipitation mechanism of SiC.

(5) The dense SiC hot-pressed at 1950°C showed the average flexural strength of 719 MPa, the Weibull modulus of 6.7, the fracture toughness of 5.0-7.8 MPa·m$^{1/2}$ and 20 GPa of Vickers hardness.

REFERENCES

[1] Y. Hirata, K. Miyano, S. Sameshima and Y. Kamino, "Reaction between SiC Surface and Aqueous Solutions Containing Al Ions", Colloid and Surfaces, A : Physicochem. & Eng. Aspects, , 133, 183-189 (1998).

[2] Y. Hirata, S. Tabata and J. Ideue, "Interaction of the Silicon Carbide -Polyacrylic Acid-Yttrium Ions System", J. Am. Ceram. Soc., 86 [1], 5-11 (2003).

[3] S. Sameshima, K. Miyano and Y. Hirata, "Sinterability of SiC Powder Coated Uniformly with Al Ions", J. Mater. Res., 13, 816-820 (1998).

[4] N. Hidaka, Y. Hirata and S. Sameshima, "Colloidal Processing of the SiC-Al$_2$O$_3$-Y^{3+} ions System and Sintering Behavior of the Consolidated Powder Compacts", J. Ceram. Proc. Res., 3 [4], 271-277 (2002).

[5] L. M. Wang and W. C. Wei, "Colloidal Processing and Liquid - Phase Sintering of SiC," J. Ceram. Soc. Jpn., 103 [5], 434 – 443 (1995).

[6] M. A. Mulla and V. D. Krstic, "Low - Temperature Pressureless Sintering of β- Silicon Carbide with Aluminum Oxide and Yttrium Oxide Additions," Am.

Ceram. Soc. Bull., 70 [3], 439 – 443 (1991).

7   G. Magnani, G. L. Minoccarri and L. Pilotti, "Flexural Strength and Toughness of Liquid Phase Sintered Silicon Carbide," Ceram. Inter., 26, 495 – 500 (2000).

8   J. H. She and K. Ueno, "Effect of Additive Content on Liquid - Phase Sintering on Silicon Carbide Ceramics," Mater. Res. Bull., 34 [10 – 11], 1629 – 1636 (1999).

9   Y. Hirata and K. Hidaka, "Surface Characteristics and Colloidal Processing of Silicon Carbide," pp.264 – 272 in Proceedings of International Symposium on Environmental Issues of Ceramics. Edited by H. Yanagida and M. Yoshimura. The Ceramic Society of Japan, Tokyo, 1995.

10   Y. Hirata and W. H. Shih, "Colloidal Processing of Two–component Powder System", pp. 637-644 in Advances in Science and Technology 14, Proceedings of 9 th Cimtec-World Ceramics Congress, Ceramics : Getting into the 2000's–Part B, Edited by P. Vincenzini, Techna Srl., Faenza, 1999.

11   S. Tabata, S. Sameshima, U. Paik and Y. Hirata, "Interaction between SiC Surface and Y^{3+} ions," J. Ceram. Proc. Res., 3 [1], 29 – 33 (2002).

12   D. J. Shaw, pp. 127 – 147 in Introduction to Colloid and Surface Chemistry, Butterworths, London, 1983.

13   L. Bergström, "Surface Chemical Characterization of Ceramic Powders", pp.71-125 in Surface and Colloid Chemistry in Advanced Ceramics Processing. Edited by R. J. Pugh and L. Bergström. Dekker, New York, 1994.

14   E. M. Levin, C. R. Robbins and H. F. McMurdie, Fig. 2388 in Phase Diagram for Ceramists, Am. Ceram. Soc., Columbus, Ohio, 1969.

15   C. F. Baes Jr. and R. E. Mesuner, p. 134 in The Hydrolysis of Cations, Robert E. Krieger Pub., Florida, 1986.

16   Y. Hirata, J. Kamikakimoto, A. Nishimoto and Y. Ishihara, "Interaction between α-Alumina Surface and Polyacrylic Acid," J. Ceram. Soc. Jpn., 100 [1], 7 - 12 (1992).

17   W. D. Kingery, "Densification During in the Presence of a Liquid Phase. I. Theory," J. Appl. Phys., 30, 301-306 (1959).

18   W. D. Kingery and M. Berg, "Study of the Initial Stage of Sintering Solid by Viscous Flow, Evaporation-condensation, and Self-diffusion", J. Appl. Phys., 26, 1205-1212 (1956).

19   C. Denoual and F. Hild, "Dynamic Fragmentation of Brittle Solids : a Multi-scale Model," Eur. J. Mechanics, 21, 105-120 (1995).

20   E. Trentini, J. Kübler and V. M. Sglavo, "Comparison of the Sandwiched Beam and Opposite Roller Loading Techniques for the Pre-cracking of Brittle Materials," J. Eur. Ceram. Soc., 23 [8] 1257-1262 (2003).

21   J. Gong, J. Wu and Z. Guan, "Examination of the Indentation Size Effect in Low-load Vickers Hardness Testing of Ceramics," J. Eur. Ceram. Soc., 19 [15], 2625-2631 (1999).

ADSORPTION OF POLY(ACRYLIC ACID) ON COMMERCIAL BALL CLAY

Ungsoo Kim, Brett M. Schulz and William M. Carty
Whiteware Research Center
New York State College of Ceramics at Alfred University
2 Pine Street; Alfred, NY 14802

ABSTRACT

The adsorption of poly(acrylic acid) on the surface of commercial ball clay was studied. The adsorption isotherm of PAAs with different molecular weights was determined at pH 6 and 9 using a titration method. The results show that the amount of PAA adsorbed on clay is independent of molecular weight. Also, the adsorption amount decreases with increased pH. However, the overall adsorption amount was lower than the range predicted by a model based upon the mineralogy of kaolinite. It was noticed that higher additions of PAA resulted in a dark supernatant during the sample preparation for titration. These dissolved species are proposed to be humic and fulvic acids, and interfered with the titration technique to determine the amount of PAA adsorbed to the clay particles. Samples of clay were washed at pH 9.5 to remove these species for identification and characterization and the adsorption of PAA on the refined clay was determined. With repeated washings the adsorption levels approach the levels predicted using the proposed model.

INTRODUCTION

Dispersion and flocculation of concentrated suspensions can be controlled through polymer adsorption and conformation. Polyelectrolytes are commonly used as dispersants of ceramic powders in aqueous media. The polyelectrolyte adsorbs at the solid-liquid interface and provides an electrostatic repulsive force between the particles, which keeps the particles well dispersed. Polyelectrolyte adsorption is highly dependent on the electrostatic interactions between the polyelectrolyte and the surface.

This work focuses on the adsorption of sodium salt of poly(acrylic acid) (Na-PAA) on the surface of commercially available ball clay. Based upon the work of Cesarano, Aksay and Bleier[1] and the clay mineralogy model[2], the adsorption levels for Na-PAA on the surface of kaolinite are predicted. Kaolinite is the major phase in ball clay and a kaolinite particle is composed of a silica-like and an alumina-like surface with edges comprised of a mixture of alumina and silica sites. The behavior of each surface in water is assumed to be similar to silica and alumina, respectively. Thus, depending on the aspect ratio of the clay platelet, the relative adsorption on the clay surface can be calculated as a percentage of the amount that would adsorb on an alumina surface.[2] The obtained adsorption isotherms are compared to the predicted level.

The purpose of this study is to verify the effect of solution chemistry and molecular weight on the adsorption behavior and to determine the effects of mineral and organic impurities, present in the raw clay, on the adsorption of Na-PAA. The results of this study are used to validate the proposed model for polymeric adsorption on the clay surface.

EXPERIMENTAL PROCEDURES
Materials

Huntingdon ball clay (United Clays, Brentwood, TN) was used in this investigation. Table I lists the chemical composition and physical properties of the clay. Surface area measurement was obtained via nitrogen adsorption (Gemini 236, Micromeritics, Norcross, GA). Na-PAAs having different molecular weights were provided. The properties of these polymers are listed in Table II.

Table I. Chemical Composition and Physical Properties of Huntingdon Clay

	SiO_2	Al_2O_3	Fe_2O_3	TiO_2	CaO	MgO	Na_2O	K_2O	LOI	Surface Area (m^2/g)
Weight percent	44.7	38.3	0.6	2.4	0.1	0.1	0.1	0.1	13.6	19.87

Table II. Reported Properties of Tested Na-PAAs

Dispersants	% Solid	M.W. (g/mol)	PDI[*]	Sources
Acumer 1010	44	2000	1.27-1.3	Rohm and Haas
Acumer 9400	42	3600	1.27-1.6	Rohm and Haas
Acusol 445N	45	4500	1.27-1.3	Rohm and Haas
Acusol 410N	40	10000	1.27-1.3	Rohm and Haas
Acumer 1510N	27	55000	1.27-1.3	Rohm and Haas
Darvan 811	43	3500-6000	1.5	R. T. Vanderbilt

[*] PDI : Polydispersity Index

Polymer Adsorption

Adsorption isotherms of Na-PAA on clay particles were determined by the solution depletion method.[3] Suspensions at 15v/o were prepared by mixing clay in de-ionized water with increasing polymer concentration from 0 to 0.6 mg/m based on clay surface area. The suspensions were adjusted to target pH values of 6 and 9 with tolerance of ±0.2. The suspensions were mixed in a shaker table for 24 hours. Suspensions were then centrifuged at 5000 rpm for 60 minutes, and the supernatant was removed without disturbing the sediment. The supernatants were adjusted to greater than pH 10 using NaOH and then titrated with 0.25N HCl to pH 2. This determines the concentration of COO⁻ groups in solution. The amount of PAA in solution was determined by generating calibration curves for PAA. The adsorption isotherm of Na-PAA (Darvan 811) on alumina (APA-0.5, Ceralox Corp, Tucson, AZ)) was also obtained to provide basis for prediction of Na-PAA adsorption on clay surfaces.

Washing and Beneficiation of the Ball Clay

Fifty pound samples of raw clay were dispersed in fifty liters of de-ionized water to create an approximately twenty volume percent suspension. The pH of the suspension was adjusted above 9.0, using 10N NaOH, to disperse the clay and dissolve any soluble organics present on the clay surface. The system was mixed for 45 minutes using a high intensity mixer (SHAR INC., Fort Wayne, IN).

After 45 minutes the mixer was turned off and the suspension was allowed to settle for a period of ninety minutes. The material remaining in suspension was decanted off and allowed to settle for an additional ten days; the settled material was named as the coarse fraction of the raw clay. After ten days the material still remaining in suspension was decanted off and dried as the fine fraction of the raw clay for characterization. The settled material was again suspended in fifty liters of de-ionized water with the pH adjusted above 9.5. This suspension was allowed to settle for an additional ten days. The material remaining in suspension was discarded and the settled material was dried as the middle fraction, assumed to be beneficiated kaolinite, for characterization.

After the beneficiation process sufficient impurity species (both organic and inorganic) remained in the clay fractions to cause deviations from the predicted adsorption levels. A process of repeated washing was used to further beneficiate the middle clay fraction for testing of polymer adsorption levels. The middle clay fraction was chosen because this is assumed to be primarily beneficiated kaolinite. A sample of the clay fraction was suspended in de-ionized water at pH above 9.5 and placed on a shaker table. The suspension was mixed for 24 hours followed by centrifuging the suspension. The material remaining in suspension was discarded and the sediment was resuspended in de-ionized water. This process was repeated until the supernatant was clear after centrifuging. The clay was then dried and used to determine the adsorption levels.

RESULTS AND DISCUSSSIONS

Effects of Suspension pH and Molecular Weight of Na-PAA on Adsorption

Figures 1 and 2 show the adsorption behavior of Na-PAA at pH 6 and 9. The adsorption plateau for tested PAAs ranges from 0.03 to 0.08 mg/m^2 at pH 9, and

from 0.1 to 0.16 mg/m² at pH 6.   All the maximum adsorption amounts are
summarized in Table III.

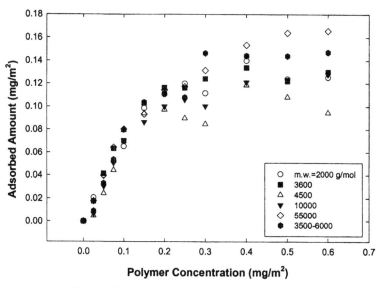

Figure 1. Adsorption isotherm of various Na-PAAs on Huntingdon clay at pH 6.
Little effect of polymer molecular weight is seen on the adsorption plateau value.

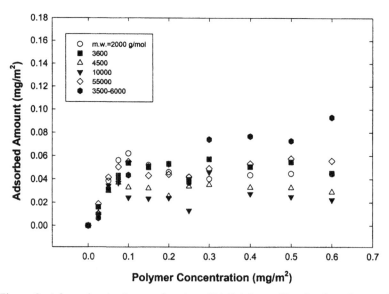

Figure 2. Adsorption isotherm of various NA-PAAs on Huntingdon clay at pH 9.

Table III. Maximum Adsorbed Amount of PAAs at Different pH of Suspensions.

	M.W. (g/mol)	pH 6 (mg/m^2)	pH 9 (mg/m^2)
Acumer 1010	2000	0.13	0.05
Acumer 9400	3600	0.13	0.05
Acusol 445N	4500	0.10	0.03
Acusol 410N	10000	0.12	0.03
Acumer 1510N	55000	0.16	0.05
Darvan 811	3500-6000	0.14	0.08

At both pH values the adsorbed amount does not essentially depend on the molecular weight. Blaakkmeer et al. developed adsorption model of weak polyacids at a charged surface in a low ionic strength medium. They expected a weak dependence on chain length at slightly higher pH values than pK_0 value and no dependence at high pH values.[4]

As pH increases, the net positive surface charge on alumina-like surface in kaolinite decreases while Na-PAA is fully dissociated and negatively charged. Therefore, adsorption of Na-PAA decreases.[1]

The adsorption behavior of Na-PAA on silica-like surface may be assumed from the Na-PAA adsorption study on silica. Small amounts of Na-PAA adsorb on silica in the acidic condition. However, Na-PAA is completely desorbed from the surface with increasing negative surface charge and increasing dissociation.[5] Therefore, in the suspensions tested here the adsorption of Na-PAA on silica-like surface will be negligible.

Carty proposed that polymer adsorption occurs on a basal surface rather than the edge based on the relative adsorption limits and SEM analysis.[2] Na-PAA has a very low affinity for silica and it is therefore assumed that there is a low affinity for the silica-like layer in the kaolinite platelet. Therefore, polymers will adsorb on the alumina-like surface as well as the edge of the clay platelet. Depending on the aspect ratio of the clay platelet, the relative adsorption (normalized in terms of ng/m^2) on the clay surface can be calculated as a percentage of the amount that would adsorb on an alumina surface. Assuming an aspect ratio of 10:1, determined to be representative from SEM evaluation of several clays, for the kaolinite platelet it can be calculated that the relative adsorption would be 41%, or adsorption on only one basal plane, to 59%, for adsorption on one basal plane plus the platelet edge. However, the experimental value was much lower than the expected values. The discrepancy was greater at pH 9.

Verification of pH Active Species in Supernatant

It is proposed that the lower adsorption amount is caused by organic impurities from ball clay, which are indistinguishable from Na-PAA in the supernatant by titration method. Prior to commencing a study to determine the effects of removing impurity species it is necessary to verify that pH active species are present in the supernatant from the clay suspension. Thus, supernatant is obtained from a Huntingdon clay suspension for analysis. A suspension was prepared at high pH, greater than 9.5, without addition of polymeric dispersant and placed on a shaker table for seven days to reach equilibrium; the pH was checked daily and maintained above 9.5. The suspension was then centrifuged and the supernatant, which was very dark, was removed and titrated using the technique described earlier. The titration curve from the supernatant is compared to that of de-ionized in Figure 3. While both curves show two inflection points, more HCl is necessary to titrate the supernatant sample relative to de-ionized water, corresponding to a concentration of 0.134 millimoles of Na-PAA in the supernatant; based upon the specific surface area of the Huntingdon ball clay this corresponds to a surface coverage of 0.041 mg/m^2. This indicates that pH active species are present in the supernatant of the clay suspension. A precipitate is formed upon titration of the supernatant.

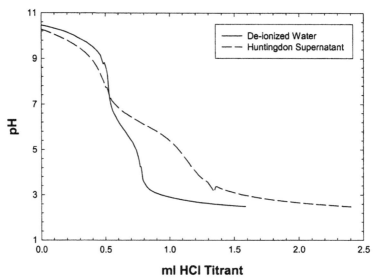

Figure 3. Titration curves for de-ionized water and supernatant from Huntingdon clay suspension.

Adsorption Isotherm on Washed Huntingdon Ball Clay

X-ray analysis was performed on each fraction of Huntingdon ball clay to see whether washing and beneficiation processes changed the mineralogy in the clay. In Figure 4 washing of the middle clay fraction is seen to have little effect on the mineral species present.

The specific surface area (SSA) of the raw clay as well as each clay fraction shown in Table IV. The SSA of the fine fraction is significantly higher than that of the coarser fractions. The fine fraction of the raw clay is mostly comprised 2:1 layer silicates, i.e. minerals that consist of an alumina-like layer sandwich between two silica-like layers, which typically have a higher SSA. It is noted that SSA of middle fraction decreases after the repeated washing. For each washing process the middle fraction is diluted, and then the trapped fine particles will be removed with each process. This results in the decrease of SSA of middle fraction after washing. However, the removal of fine particles increases the adsorption amount of Na-PAA because Na-PAA has very low adsorption affinity for fine particles, which are mainly 2:1 layer silicates.

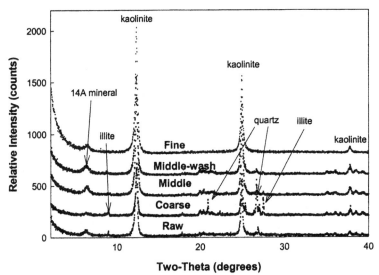

Figure 4. X-ray diffraction patterns from oriented slides of Huntingdon ball clay and the clay fractions prepared by washing and beneficiation. Washing of the middle clay fraction is seen to have little effect on the mineral species present.

Table IV. Measured specific surface area of each fraction from Huntingdon ball clay and middle fraction after repeated washing.

	Raw	Coarse	Middle	Middle-wash	Fine
Specific surface area (m²/g)	19.87	10.71	10.26	9.19	32.08

The adsorption isotherms are shown for pH 6 (Figure 5) and pH 9 (Figure 6). Only the raw clay and middle fraction were tested for the Huntingdon ball clay. There is little difference in the adsorption isotherms for the raw clay and the middle fraction at pH 6. After successive washing of the middle fraction the plateau value increases significantly due to the removal of impurity species prior to the adsorption study and the adsorption level is seen to approach that predicted by the model of the kaolinite platelet. Weak adsorption of the polymer on the clay is seen in all of the isotherms at pH 9, but all of the samples are seen to reach approximately the same plateau value with little benefit of successive washing of the middle fraction. This is due to the soluble organics remaining in the supernatant after centrifuging from all of the suspensions that were prepared. The detected concentration of organic in the supernatant with no polymer addition was taken as a "background" concentration and subtracted from the result prior to analysis. In these cases at pH 9 the adsorption level is seen to approach that predicted by the model of the kaolinite platelet.

From the above observations it is shown that the discrepancy in absorption amount is caused by the interference of organic impurities in the clay. It is known that organic matter exists in most mineral soils in the form of humic acid (HA) and Fulvic acid (FA).[6] Schulthess and Huang showed that HA and FA strongly adsorbed on kaolinite from pH 2 to 10. They determined that adsorption occurred mostly on the alumina-like surface of kaolinite by comparing with the adsorption on silica and alumina.[7] Jones and Bryan also reported that the adsorption of HA decreases with increase in pH.[8] From these it can be concluded that Na-PAA should compete with pre-absorbed HA and FA molecules for adsorption. This potentially explains the low adsorption values for Na-PAA. At pH 9 there will be a very limited extent of adsorption sites on kaolinite particles. Therefore, the competition between Na-PAA and HA and FA for adsorption will be strong. This explains the higher discrepancy at pH 9. In addition, carboxyl groups on dissolved HA and FA interact with OH⁻ during titration of supernatant, which results in the overestimation of the amount of unadsorbed Na-PAA. This leads to the low adsorption amounts determined experimentally for Na-PAA.

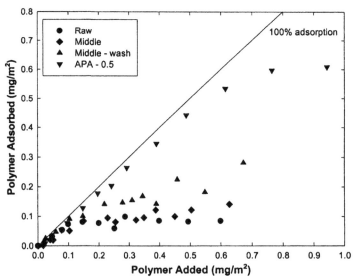

Figure 5. Adsorption isotherms for Huntingdon ball clay and alumina (APA-0.5) at pH 6. After repeated washing of the middle fraction water the plateau value approaches that predicted by the model of the kaolinite platelet.

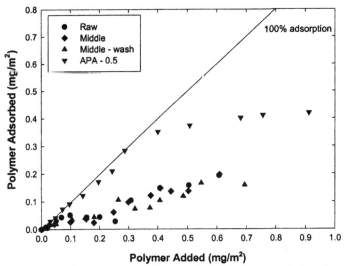

Figure 6. Adsorption isotherms for Huntingdon ball clay and alumina (APA-0.5) at pH 9. Little effect of the washing and beneficiation process is seen in the adsorption plateau values. All of the clay fractions reach approximately the same plateau value at higher polymer additions. The concentration of soluble organic species present in the suspensions prepared with no polymer addition was taken as a background reading and subtracted from the concentration of polymer detected in the supernatant.

CONCLUSIONS

This study confirms the adsorption behavior of Na-PAA. The adsorption of Na-PAA on clay particles decreases with increase in pH. The surface charge of the clay particle and the polymer conformation change due to polymer charge explain the adsorption behaviors of Na-PAA on clay particles. The Na-PAA adsorption amount is not influenced by the molecular weight. The obtained Na-PAA adsorption isotherms confirm the proposed adsorption model on clay particle, in which polymer adsorption occurs on a basal surface rather than the edge of clay platelet. The increased adsorption amount on middle fraction after washing again proves that the polymer adsorption occurs on basal surface. The increased adsorption occurs on the newly exposed basal surface. Otherwise the adsorption amount should not change after the washing process.

References
. Cesarano III, I. A. Aksay, and A. Bleir, "Stability of Aqueous α-Al₂O₃ Suspensions with Poly(methacrylic acid) Polyelectrolyte," *J. Am. Ceram. Soc.*, **71** [4] 250-255 (1998).
. W. M. Carty, "The Colloidal Nature of Kaolinite," *Am. Ceram. Soc. Bull.*, **78** [8] 72-76 (1999).

3. G. J. Fleer, M. A. Cohen Stuart, J. M. H. M. Scheutjens, T. Cosgrove, and B. Vincent, *Polymers at Interfaces*; pp. 17-24, 339-371. Chapman & Hall, London, UK, 1998.
4. J. Blaakkmeer, M. R. Bohmer, M. A. Cohen Stuart, and G. J. Fleer, "Adsorption of Weak Polyelectrolytes on Highly Charged Surfaces. Poly(acrylic acid) on Polystyrene latex with Strong Cationic Groups," *Macromolecules*, **23** [8] 2301-2309 (1990).
5. G. R. Jopplen, "Characterization of Adsorbed Polymers at the Charged Silica-Aqueous Electrolyte Interface," *J. Phys. Chem.*, **82** [20] 2210-2215 (1978).
6. T. B. S. Christopher, "Humic Substances," (1996) Universiti Putra Malaysia. Accessed on: September 2002. Available at <http://agri.upm.edu.my/jst/resources/as/om_humicsubs.html>
7. C. P. Schulthess and C. P. Huang, "Humic and Fulvic Acid Adsorption by Silicon and Aluminum Oxide Surfaces on Clay Minerals," *Soil Sci. Soc. Am. J.*, **55** [1] 34-42 (1991).
8. M. N. Jones and N. D. Bryan, "Colloidal Properties of Humic Substances," *Adv. Colloids Interface Sci.*, **78** [1] 1-48 (1998).

ANALYSIS OF THICKNESS CONTROL VARIABLES IN TAPE CASTING
PART II : THE EFFECT OF BLADE GAP

Matthew C. Gifford
Alfred University
Alfred, NY 14802

Eric R. Twiname
Richard E. Mistler, Inc.
1430 Bristol Pike
Morrisville, PA 19067

Richard E. Mistler
Richard E. Mistler, Inc.
1430 Bristol Pike
Morrisville, PA 19067

ABSTRACT

Several models have been previously proposed to explain the formation of tape thickness in horizontal bed, doctor blade tape casting[1,2,3,4]. The complexity, or sheer number of influential variables in the casting system, however, has thus far prohibited the emergence of a single model that serves to mathematically represent the entire system. The reported work contains part of an ongoing empirical study of wet and dry tape thicknesses and the influence of two basic control variables, carrier speed and doctor blade gap, on these thicknesses in order to more thoroughly understand the effect of process parameters. Dry tape thickness was measured while wet tape thickness was calculated using a conversion equation, reported within, based solely upon measurable quantities.

INTRODUCTION

The thickness of tape cast layers plays an important role in the performance of many ceramic and non-ceramic applications. For example, ceramic capacitors and thermistors rely on layer thickness control, among other variables, to determine proper device performance. Ceramic packages also rely on layer thickness to provide adequate electrical insulation, signal isolation and, in some cases, proper spacing for capacitor electrodes. Tape cast layer thicknesses are also controlled in various other applications to achieve: desired wear life, weight, oxygen diffusion rate, strength, or simply material mass for brazing or sheet metal forming or further shaping processes.

While various theories and methods exist regarding the formation, prediction and control of tape thickness, a unifying theory for tape cast thickness control has yet to emerge. The generally accepted flow model[1] describes flow under the doctor blade as a sum of Pressure Flow through an orifice and Couette Flow stemming from the moving carrier film. A quantitative exploration of thickness control variables and their effect on tape thickness will lead to a more thorough understanding of not only the casting process, but will also aid in differentiating which control variables have major or minor effects for various target thicknesses. Isolation of major vs. minor control variables will aid in the efficient application of time, money and other resources towards thickness control. Toward that end, a single slip was chosen around which to build a quantitative database of control variable changes and the resulting effects on wet and dry thicknesses.

EXPERIMENTAL PROCEDURE:
Casting Slip

The slip formulation[5,6] used for this study is a standard 95% alumina body doped with clay and talc formulated using a xylenes/ethanol solvent blend, menhaden fish oil dispersant, polyvinyl butyral binder and butyl benzyl phthalate plasticizer. Each batch of slip is prepared by dispersion milling the solvent, dispersant and powders for 22-26 hours in a size O glass fortified jar mill ¼ filled with ½" cylindrical 96% alumina media. Plasticizer and binder are then added to the mill and rolled for an additional 22-26 hours.

Each slip batch is deaired under 25 inches of mercury vacuum for eight (8) minutes with agitation. Single point viscosity (RV4@20rpm) is then measured using a Brookfield rotational viscometer and the slip temperature recorded. Average viscosity for the slips used to date was 1600 cP. Specific Gravity (Sp. Grav.) is measured using 100 ml slip samples in a graduated cylinder. Specific Gravity for this slip (after deairing) is found to be 1.67 g/cc.

Tape Casting

The prepared slip was cast onto silicone coated PET using a twenty five foot horizontal bed tape casting machine. Laydown was done using a stationary 0.25 inch flat bottomed doctor blade assembly riding directly on the carrier film, supported by a granite surface plate (Fig. 1). Carrier speed (Vc) was controlled using a closed loop feedback system. Reservoir height was held constant at 1.25 inches by hand addition of slip to the reservoir throughout the casting process. Each prepared slip batch was cast into a single tape at a single blade gap (G). Tapes were cast using blade gaps of 0.005 inch (5 mil, 125 micron) through 0.030 inch (30 mil, 750 micron) in 5 mil increments. The carrier speed was varied during each cast to achieve a minimum of 2 feet of steady state tape per gap-speed combination. Speeds used include 5 through 30 inch/minute (ipm) at 2.5 ipm increments and 36, 42, 48 ipm. Each cast was performed, and dried, under identical airflows and ambient temperatures. All tapes were allowed to dry for at least 24 hours prior to removal from the machine.

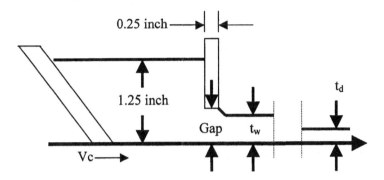

FIGURE 1: Schematic of Tape Casting Process Parameters Studied

Analysis

Dry thickness data (t_d) was collected using a 0.25 inch diameter flat faced micrometer. Measurements were recorded at initial resistance to micrometer barrel motion to avoid compression of the tape and resulting measurement errors. Green bulk density (G.B.D.) of 1 x 1.5 inch precision punched blanks (25 x 37 mm) was calculated by measuring the thickness of the blank in five (5) places and measuring the weight of the blank. A minimum of 100 discrete thickness measurements and 20 discrete green bulk density measurements were made for each gap-speed combination and plotted to ensure that no error trends such as streaks or cyclic variation existed. Wet thickness data (t_w) for each gap-speed combination was calculated via Equation 1, using the corresponding dry thickness data, density and specific gravity for that tape section.

$$t_w = t_d \cdot \left(\frac{G.B.D.}{Sp.Grav.} \right) \cdot \left(\frac{Mass_{Slip}}{Mass_{tape}} \right) \tag{1}$$

The only assumption made in the derivation of this equation is that the cast wet film does not shrink laterally during drying. This assumption permits the volume component of the Green Bulk Density to Specific Gravity ratio to be used as a thickness ratio. The following mass ratio, calculated from the slip batch formula used, accounts for the loss of solvent mass during drying. A simple variant of Equation 1, shown as Equation 2, was used to calculate the thickness shrinkage during drying for each gap-speed combination.

$$\frac{t_w}{t_d} = \left(\frac{G.B.D.}{Sp.Grav.} \right) \cdot \left(\frac{Mass_{Slip}}{Mass_{tape}} \right) \qquad (2)$$

It should be specifically noted that the thickness of the wet cast (t_w) cannot be used interchangeably with doctor blade gap in these equations as will be shown in Figures 3-6 below.

RESULTS AND DISCUSSION
Wet to Dry thickness Ratio
    The shrinkage ratio during drying was examined for all of the tapes and found to be very consistent (Fig. 2), ranging from 2.01 to 2.11 with no discernable trends. Since the calculated values for this ratio stem from Eqn. 2, it also holds that very little variation was measured in green bulk density though dry tape thicknesses ranged from less than 5 to over 28 mils (125-710 microns). While the wet/dry thickness ratio was found to be approximately 2:1 for this slip, other slips not reported here have had significantly different shrinkage ratios.

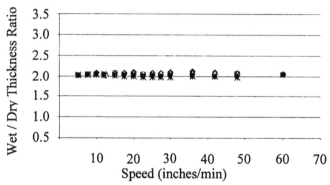

FIGURE 2: Wet/Dry Shrinkage During Drying
for Blade Gaps from 0.005 inch – 0.030 inch

The Effect of Carrier Speed
    The measured wet and dry thickness values obtained are shown in Figures 3 and 4. Since the difference between wet and dry thickness values in this work is fairly constant, and the intent of this work is to understand slip behavior during casting, only the wet thickness values will be handled in discussions below. It is generally accepted, and shown in all of the plotted data herein, that tape thickness decreases with increasing carrier speed.

As can be seen from Fig. 3, carrier speed has more influence on wet thickness when using larger blade gaps. Little change in wet thickness is noted for the 5 mil (125 μm) blade gap as speed is increased from 5 ipm up to 48 ipm.

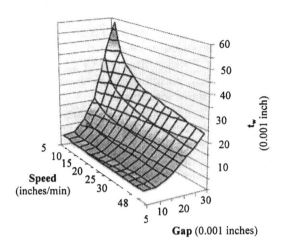

FIGURE 3: Calculated Wet Thickness as a Function of Casting Speed and Blade Gap.

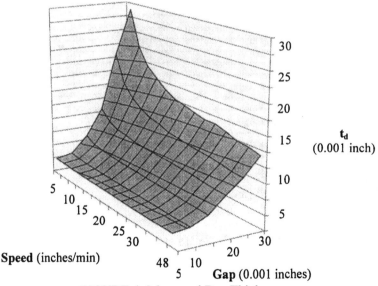

FIGURE 4: Measured Dry Thickness as a Function of Casting Speed and Blade Gap.

Tape cast at G=30, however, shows a strong dependence upon casting speed, dropping from near 60 mils at 5 ipm to less than 25 mils at 48 ipm. It should also be noted that changes in casting speed have a greater effect on thickness at low speeds than at higher casting speeds. It can be seen in Figure 5 that the major effect of casting speed was found between 5-20 ipm where wet thickness for G=30 decreased by 30 mils while increasing casting speed to 48 ipm resulted in a further thickness drop of 7 mils. Both the G=25 and G=20 curves also begin to flatten at approximately 20 ipm. The G=5 curve, in contrast, shows little change throughout the range of speeds used.

The drying time for tapes decreases significantly as tape thickness decreases. This translates, in practice, to a generally faster casting speed for thin tapes that dry quickly and a slower casting speed for thicker tapes needing more dwell time in the tape casters drying chamber. Thinner tapes, therefore, are not only less sensitive to casting speed variation but tend to be cast at higher speeds at which carrier speed variation has less effect in general. It is also interesting to note that some of the wet thickness values shown are greater than the blade gaps with which they were cast. In the case of G=30 cast at 5 ipm, the wet thickness is near double the blade gap (57.9 mil ± 2.2 full range) and the dry thickness was near equivalent to the gap itself (28.8 mil ± 1.15 full range). One standard deviation error bars are smaller than the data points shown in all but two cases and thus are not shown on the figures.

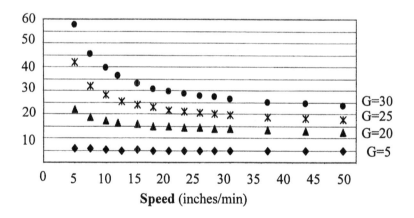

FIGURE 5: Calculated Wet Thickness as a Function of Casting Speed.

The Effect of Blade Gap

The collected data shows that tape thickness increases with increasing blade gap as would be expected. Changes in blade gap are also seen to have a greater effect at larger gaps. It can be seen in Fig. 6 that the slope of each curve

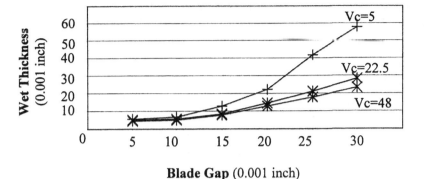

Blade Gap (0.001 inch)
FIGURE 6: Calculated Wet Thickness as a
Function of Doctor Blade Gap.

increases with increasing gap. Based upon the additive Pressure/Couette flow model, this behavior would be attributed to pressure flow becoming dominant in thickness determination as blade gap increases. Counter to this, there is little difference in wet thickness between blade gaps of 5 mil and 10 mil (125 – 250 microns), regardless of speed changes from 5 ipm to 48 ipm. While it could be assumed that Couette flow is dominant in this range, the small change in wet thickness resulting from a doubling of the blade gap does not fit well with the linear proportionality between volume flow and blade height in the Couette model, to say nothing of the cubic proportionality of the pressure model. It would appear that, with this slip, another variable not yet examined is dominant in this range of speeds and blade gaps. Since reservoir height serves only to define the pressure drop across the orifice under the blade, it is believed that slip rheology is more likely to be a causal factor in this observed behavior. The effects of slip rheology and reservoir height on wet and dry tape thicknesses are subjects of proposed future work.

CONCLUSIONS
    The work reported here, as mentioned earlier, reflects only part of an ongoing empirical analysis of thickness determining variables. This being the case, the conclusions drawn from this work are limited to the work itself. Some of the conclusions will be true for all slips while other, more specific conclusions may only hold true for the slip used in this study.
    It is shown in this study that wet and dry tape thicknesses decrease, in general, with decreasing doctor blade gap. This decrease is shown to be non-linear. Tape thickness is more sensitive to changes in blade gap at large gaps and low carrier speeds and less so with relatively small gaps or with high carrier speeds.

Tape thickness also is seen to decrease with increasing casting speed. The speed dependence measured was also found to be non-linear, with greater speed dependence at lower casting speeds and large blade gaps, and relatively little effect at high speeds or with smaller blade gaps. A method for calculating wet tape thickness using only measurable and/or verifiable quantities is given in this work.

REFERENCES-
[1] Y.T.Chou, Y.T.Ko and M.F.Yan, "Fluid Flow Model for Ceramic Tape Casting," *Journal of the American Ceramic Society*, **70** [10] C280-C282 (1987).

[2] T.A.Ring, "A Model of Tapecasting Bingham Plastic and Newtonian Fluids," pp. 569-576 in Advances in Ceramics, Vol. 26, Edited by M.F.Yan et.al. (1989).

[3] R.Pitchumani and V.M.Karbhari, "Generalized Fluid Flow Model for Ceramic Tape Casting," *Journal of the American Ceramic Society*, **78** [9] 2497-2503 (1995).

[4] G.Terrones, P.A.Smith, T.R.Armstrong and T.J.Soltesz, "Application of the Carreau Model to Tpae-Casting Fluid Mechanics," *Journal of the American Ceramic Society*, **80** [12] 3151-3156 (1997).

[5] R.E.Mistler and E.R.Twiname, *Tape Casting: Theory and Practice*, The American Ceramic Society, Westerville, OH, 2000.

[6] R. Cooley, E.R. Twiname, R.E. Mistler, "Analysis of Thickness Control Variables in Tape Casting – Part I," 2002 Annual Meeting of the American Ceramic Society, St. Louis, MO, April 30, 2002.

KEYWORD AND AUTHOR INDEX

Adsorption, 1, 119, 129
Aggregated suspensions, 65
Alumina, 47
Amino acid, 1
Andersson, K.M., 93
Anionic polymer dispersant, 83
Atomic Force Microscopy (AFM), 83, 93

Ball clay, 129
Barium titanate, 27
Bergstrom, L., 93
Blade gap, 139

Calcination, 55
Capacitor, 27
Carty, W., 129
Ceria, 1
Colloidal coating, 55
Colloidal probe AFM, 83
Compressive yield stress, 65
Crystal, 11
Curing, 37

Dispersant, 93
Dispersion, 1
Dissolution, 93
Doctor blade, 139

Electro-photographic solid freeform
fabrication, 75
Ferroelectric, 11

Flexural strength, 119
Fracture toughness, 109
Friction, 93

Gas separation, 47
Gifford, M., 139
Glass-ceramic, 11
Gomi, K., 27
Gu, H., 55

Hard metal, 93
Heterocoagulation, 109
Hidaka, N., 109
Hirata, Y., 1, 109, 119
Hot-pressing, 119

Iida, Y., 27

Kakui, T., 83
Kamiya, H., 27, 83
Kaolinite, 129
Kim, U., 129

Lead magnesium niobate, 55
Lee, S., 75
Liquid phase sintering, 109

Manjooran, N., 75
Matsui, S., 83
Micromechanical testing, 65
Microporous, 47
Miller, K.T., 65
Mistler, R., 139
Mutharasan, R., 47

Na-PAA, 129
Nucleation, 11

Ogino, K., 27
Oligomer, 27
Optical density, 75
Organic matter, 129

Patil, T., 47
Perovskite, 11, 55
Phase separation, 1
PMN, 55
PMN-PT, 55
Polyacrylic acid, 119
Polystyrene, 65
Porosity, 93
Promkotra, S., 65
Pyrgiotakis, G., 75

Rapid prototyping, 75
Rheology, 75

Saegusa, K., 11
Sameshima, S., 1
Schulz, B., 129
Shih, W.-H., 37, 47, 55
Shih, W.Y., 37, 47, 55
Shimazu, H., 1
Sigmund, W., 75
Silica, 47
Silicon carbide, 75, 109, 119
Silicon nitride, 83
Sintering additive, 109

Tabata, S., 119
Takahashi, H., 1
Tanaka, K., 27
Tape, 139
Tape casting, 139
Thermal stability, 37
Thickness, 139
Thickness control, 139
Thin film, 11
Titration, 129
Tungsten carbide-cobalt, 93
Twiname, E., 139
Two-dimensional colloids, 65

Yonemochi, Y., 27
Yttrium ion, 119

Zeta potential, 1
Zhao, Q., 37, 47
Zirconia, 37